SpringerBriefs in Electrical and Computer Engineering

More information about this series at http://www.springer.com/series/10059

Yuanyuan Yang • Cong Wang

Wireless Rechargeable Sensor Networks

 Springer

Yuanyuan Yang
Department of Electrical
 and Computer Engineering
Stony Brook University
Stony Brook, NY, USA

Cong Wang
Department of Electrical
 and Computer Engineering
Stony Brook University
Stony Brook, NY, USA

ISSN 2191-8112   ISSN 2191-8120   (electronic)
SpringerBriefs in Electrical and Computer Engineering
ISBN 978-3-319-17655-0   ISBN 978-3-319-17656-7   (eBook)
DOI 10.1007/978-3-319-17656-7

Library of Congress Control Number: 2015936334

Springer Cham Heidelberg New York Dordrecht London

Printed on acid-free paper

Springer International Publishing AG Switzerland is part of Springer Science+Business Media (www.
springer.com)

Recommended by Xuemin (Sherman) Shen

Preface

With an ever increasing demand of new wireless sensing applications, energy has been the primary concern for wireless sensor networks. In particular, energy conservation has been studied extensively to extend network lifetime. A variety of approaches have been proposed in literature that can elongate network lifetime to some extent. However, with limited energy storage, sensor's battery would deplete eventually and replacing those batteries requires tremendous human efforts.

In this book, we describe a new approach to replenishing sensor's battery via wireless charging without wires or plugs. We start with a detailed overview of the recent developments in wireless charging technologies and their applications in wireless sensor networks to highlight the advantages and disadvantages. We then provide a new hierarchical network architecture that adopts a mobile vehicle for wireless charging. We name such networks *Wireless Rechargeable Sensor Networks (WRSNs)*. Based on this network architecture, we discuss several principles from theoretical aspects, and design communication protocols and recharge scheduling algorithms to maintain perpetual network operations. We also give network performance evaluation results in various criteria such as nonfunctional node percentage, network latency, and energy overhead. The state-of-the-art wireless charging technology covered in this book would help readers understand existing challenges and inspire future research to improve energy efficiency and network lifetime for wireless sensor networks.

Stony Brook, NY, USA   Yuanyuan Yang
February 2015   Cong Wang

Acknowledgments

The writing of this book was supported in part by the grant from US National Science Foundation under grant number ECCS-1307576.

Contents

Acronyms

AC   Alternating Current
DC   Direct Current
CMST   Capacitated Minimum Spanning Tree
CVRP   Capacitated Vehicle Routing Problem
EIRP   Effective Isotropic Radiated Power
EV   Electrical Vehicle
EW   Esau-Williams
FCC   Federal Communication Commission
TSP   Traveling Salesman Problem
VRP   Vehicle Routing Problem
VRPTW   Vehicle Routing Problem with Time Windows
WRSN   Wireless Rechargeable Sensor Network
WSN   Wireless Sensor Network

Chapter 1
Introduction

1.1 Introduction and Background

The next generation wireless networks rely on sensors to identify and extract useful information from the environment. With the option to mount various types of detectors ranging from temperature, magnetic, pressure, acoustic sensors to more complex gyroscope, imaging, infrared, video sensors, wireless sensor networks (WSNs) provide an easy way to access information in the physical world [1, 2]. It begins to find an increasing number of applications from our daily life to many mission-critical tasks. Typical examples in our daily life include temperature and humidity sensors deployed indoors that can automatically control the climate. In mission-critical tasks such as volcano or forest fire monitoring [3, 4], sensors also play an irreplaceable role to provide accurate readings on time. For example, the traditional forest fire monitoring system depends on the analysis of satellite images. However, the accuracy of these systems is usually limited by image quality and weather conditions. Sensors equipped with thermal imaging and temperature detectors can be deployed and transmit real-time data in a designated area [4].

The increasing demand for more complex sensors leads to higher energy consumption on sensor nodes. To this end, energy conservation has been one of the primary focuses in WSN research in the past decade. Since replacing sensor's battery is infeasible or risky in many applications [3, 4], most of the research aims to maximize network lifetime. For a single node, duty cycling is one of the most effective methods to save energy [10]. It puts radio transceivers in sleep mode whenever there is no communication. To adopt this method in a network, wakeup/sleep scheduling of sensors is required to guarantee end-to-end communications [11, 12]. In addition, battery-aware routing and scheduling based on battery recovery property have been studied to extend sensor node lifetime [13–15]. At the network level, researchers have considered maximizing network lifetime by optimizing either flow routing [16] or sensor missions [17, 18]. Besides, how data is collected also determines network lifetime. Traditional approach to

Y. Yang, C. Wang, *Wireless Rechargeable Sensor Networks*, SpringerBriefs
in Electrical and Computer Engineering, DOI 10.1007/978-3-319-17656-7_1

aggregating sensed data through a static data sink is known to be less energy efficient since nodes close to the sink consume more energy to relay packets. These nodes usually form a bottleneck around the sink and put an upper limit on the network lifetime while other nodes may still have energy. This is regarded as the infamous "energy hole problem" [19]. A solution is to introduce a mobile data sink for data gathering [20–26]. It has been shown [25] that by carefully planning trajectory of the mobile sink, energy consumptions on sensor nodes can be balanced and network lifetime can be extended significantly.

Although these methods can prolong network lifetime to some extent, sensor's battery would deplete eventually and cause service interruptions. A promising technique is to renew sensor's battery by harvesting environmental energy such as solar and wind [5–7]. For example, solar harvesting can provide energy from solar panels of similar size to sensor nodes [8]. It is also shown that multiple ambient energy sources can be utilized to power sensor nodes in [9]. However, an inevitable drawback of environmental energy harvesting is due to the inherent dynamics of energy sources. When energy sources are not available, sensor nodes may stop working and it can lead to long data latency or data loss in the network.

Recently, finding an easy and reliable way to replenish sensor's battery begins to attract more attentions in the sensor network research community. Fortunately, breakthroughs in wireless charging technology have opened up a new dimension to power sensor nodes in distance without any wires or plugs. Pioneered by Nikola Tesla [27] a century ago, it is only recently wireless charging enjoys so much popularity after the experimental realization by Kurs et al. [28]. It has been shown in [28] that a total of 60 W energy can be transferred between two magnetically coupled coils over an air gap of 2 m with 40 % efficiency. The experimental prototype is soon extended to power multiple devices in [29]. In the meanwhile, fast development of mobile devices and stagnant battery technology deliver the impetus to drive wireless charging technology into commercialization and many products are now available. For example, charging pad called "Powermat©" can recharge multiple cell phones and PDAs simultaneously by simply putting them on the pad [30]. Powercast© systems realize wireless charging for sensing devices up to several meters away [31]. This technology has demonstrated not only the strengths to power small portable devices, but also the potentials to recharge Electrical Vehicles (EVs). With the ability to deliver 100 W of energy at high efficiency, wireless charging systems can be launched at power stations, parking lots or even beneath road surface to recharge EVs without any physical contact [32].

The main focus of this book is to examine how to employ wireless charging in traditional battery-powered wireless sensor networks and we call such networks *Wireless Rechargeable Sensor Networks (WRSNs)* henceforth. We start with a detailed overview of the recent developments in wireless charging technologies and their applications in WSNs to highlight the advantages and disadvantages. We then introduce controlled mobility to a hierarchical network in order to provide efficiency and scalability. Based on the new network architecture, we discuss several important principles from theoretical aspects. We also provide a distributed communication protocol for gathering node status information in real-time, followed by recharge scheduling algorithms that aim to maintain perpetual network operations. Finally, we give network performance evaluation results in various criteria such as nonfunctional node percentage, network latency, energy overhead, etc.

1.2 Wireless Charging Technology

In this section, we introduce two major techniques of wireless charging: electromagnetic radiation and magnetic resonant coupling, and their applications in WSNs.

1.2.1 Electromagnetic Radiation

Electromagnetic waves have been used for communications since the last century. Recently, upon discovering the energy resides in the electromagnetic waves can be captured to power ultra-low power devices, a great amount of research efforts have been devoted to scavenge energy in the ubiquitous electromagnetic waves. There are plenty of such energy sources such as TV towers, cellular stations or even local Wi-Fi access points. However, due to the nature of isotropic wave propagation, received signal strength decreases dramatically with transmission distance. Thus only a very small fraction of energy can be effectively captured from the air. Figure 1.1a shows a sketch of an electromagnetic radiation based wireless charging system.

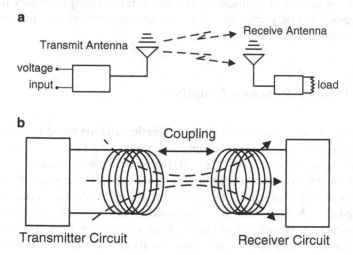

Fig. 1.1 Wireless charging systems. (**a**) Electromagnetic radiation. (**b**) Magnetic resonant coupling

A popular commercial product currently on the market is the Powercast© wireless charging system [31]. It consists of a wireless energy transmitter operating at 850–950 MHz and a few receivers. There has been some work on applying such systems in WRSNs. In [33], the impact of wireless charging on current routing and node deployment schemes in WSNs is studied. In [34], the problems of how to place and mobilize wireless chargers to sustain network operations are studied. First, a point provisioning problem is proposed to ensure any position of the network can receive enough energy. Then a path provisioning problem is studied to further reduce the number of wireless chargers. The problem is extended in [35] to minimize charging delay by optimally planning the moving trajectory of mobile chargers. In [36], an $\mathcal{O}(k^2 k!)$ (where k is the number of nodes in the network) algorithm is designed to schedule recharge activities such that network lifetime is maximized. In [37], a joint routing and wireless charging scheme is proposed by guiding routing and recharge activities. In addition, problems in WRSNs other than recharge scheduling are studied in [38, 39]. In [38], an important safety issue of using electromagnetic radiation based wireless charging is studied. Since absorption of overdosed electromagnetic radiation poses great risks to human body, a placement problem on how to place wireless chargers to sustain network operations while the radiation level of all positions is below a threshold is studied in [38]. Other than the safety issue, in [39], it is shown that traditional localization strategies in WSNs can be further improved by measuring the wireless charging time of sensors.

As pointed out in [38], a limitation of electromagnetic radiation based wireless charging is due to health concerns. Although it is desired to increase the emitted energy at the power source, the Federal Communication Commission's (FCC) has a regulation of maximum effective isotropic radiated power (EIRP) at 4 W [40]. In addition, the isotropic nature of omni-directional antenna emits energy that attenuates quickly over distance. Therefore, this technique usually has very low efficiencies and only supports low-power sensing applications such as simple temperature, humidity monitoring, etc.

1.2.2 Magnetic Resonant Coupling

In contrast to the low-efficiency in electromagnetic radiation based wireless charging techniques, magnetic resonant coupling can transfer a large amount of energy over an air gap at high efficiency [28, 29]. Figure 1.1b shows a wireless charging system with magnetic resonant coupling. To guarantee high charging efficiencies, a mobile vehicle with high-density battery packs is usually adopted to get to sensors in close distance. A transmitter coil is mounted on the vehicle to transport energy from its battery to sensors. While approaching sensors, the vehicle first converts direct current (DC) output from the battery to alternating current (AC) through a DC/AC converter to induce an oscillating magnetic field around the transmitting coil. On the sensor side, the receiving coil is tuned to resonate at the same frequency. An alternating current is generated at the sensor's output circuit. The AC is then converted back to DC to recharge sensor's battery.

The potentials of using magnetic resonant coupling in WRSNs is studied in [41–45]. In [42], an optimization problem to maximize the ratio between charging vehicle idling and working time is studied. A Hamiltonian cycle through all the sensor nodes is proved to be the shortest recharge path. Instead of recharging all the nodes, in [41], only a number of nodes request for recharge are serviced, and this number is upper bounded by a tour length threshold to guarantee data latency. During recharge, the charging vehicle simultaneously gathers data from the neighborhood in multi-hops and uploads all collected data to the base station after a recharging cycle is completed. A system-wide optimization is performed to maximize network utility by selecting optimal data rates and flow routing. In [44], optimal allocation of vehicle's stopping time to recharge sensors at different locations is studied. Upon realizing the dynamics in sensors' energy consumptions, to provide more accurate recharge decisions, a recharge framework is proposed in [45]. An NP-hard problem to minimize the movement cost of charging vehicles is studied and several heuristic algorithms are proposed. In sum, magnetic resonant coupling is a promising technology ready to support many energy-demanding multimedia applications with enormous data communication and sensing activities. Therefore, in this book, we focus on applying this technology in WSNs.

1.3 Summary

In this chapter, we have presented a brief introduction to wireless charging technology and its application in WSNs. We discuss its latest advances and describe two typical techniques to perform wireless charging followed by a literature review of the most recent works in wireless sensor research community. The subsequent chapters provide a detailed coverage of several important issues in WRSNs including the basic network architecture, components, principles, a distributed node status reporting protocol, recharge scheduling algorithms and performance evaluations.

References

1. I. Akyildiz, "Wireless sensor networks: a survey," *Computer networks*, vol. 38, no. 4, pp. 393–422, 2002.
2. K. Sohrabi, "Protocols for self-organization of a wireless sensor network," *IEEE personal communications*, vol. 7, no. 5, pp. 16–27, 2000.
3. W. A. Geoffrey, "Deploying a wireless sensor network on an active volcano," *IEEE Internet Computing*, vol. 10, no. 2, pp. 18–25, 2006.
4. L. Yu, N. Wang, X. Meng, "Real-time forest fire detection with wireless sensor networks," *IEEE International Conference on Wireless Communications, Networking and Mobile Computing*, vol.2, pp.1214–1217, 2005.

5. T. Voigt, H. Ritter, J. Schiller ,"Utilizing solar power in wireless sensor networks," *IEEE International Conference on Local Computer Networks (LCN)*, 2003.
6. M. Rahimi, H. Shah, G. Sukhatme, J. Heideman and D. Estrin, "Studying the feasibility of energy harvesting in a mobile sensor network," *IEEE International Conference on Robotics and Automation*, 2003.
7. C. Wang, S. Guo and Y. Yang, "Energy-efficient mobile data collection in energy-harvesting wireless sensor networks," *The 20th IEEE International Conference on Parallel and Distributed Systems (ICPADS 2014)*, Hsinchu, Taiwan, Dec. 2014.
8. J. Paradiso, T. Starner, "Energy scanverging for mobile and wireless electronics," *IEEE Journal of Pervasive Computing*, vol. 4, no. 1, pp. 18–27, 2005.
9. C. Park and P. H. Chou, "AmbiMax: autonomous energy harvesting platform for multi-supply wireless sensor nodes," *IEEE International Conference on Sensing, Communication, and Networking (SECON)*, vol. 1, pp. 168–177, 2006.
10. W. Ye, J. Heidemann, D. Estrin, "An energy-efficient MAC protocol for wireless sensor networks," *IEEE INFOCOM*, vol.3, pp. 1567–1576, 2002.
11. A. Keshavarzian, H. Lee and L. Venkatraman, "Wakeup scheduling in wireless sensor networks," *ACM International Symposium on Mobile Ad Hoc Networking and Computing (MobiHoc)*, pp. 322–333, 2006.
12. Z. Zhang, M. Ma and Y. Yang, "Energy-efficient multi-hop polling in clusters of two-layered heterogeneous sensor networks," *IEEE Transactions on Computers*, vol. 57, no. 2, pp. 231–245, Feb. 2008.
13. C. Ma and Y. Yang, "A battery-aware scheme for routing in wireless ad hoc networks," *IEEE Transactions on Vehicular Technology*, vol. 60, no. 8, pp. 3919–3932, Oct. 2011.
14. C. Ma, Z. Zhang and Y. Yang, "Battery-aware scheduling in wireless mesh networks," *ACM/Springer Mobile Networks & Applications (MONET)*, vol. 13, pp. 228–241, 2008.
15. C. Ma and Y. Yang, "Battery-aware routing for streaming data transmissions in wireless sensor networks," *ACM/Springer Mobile Networks & Applications (MONET)*, vol. 11, no. 5, pp. 757–767, October 2006.
16. C. J. Hwan and L. Tassiulas, "Maximum lifetime routing in wireless sensor networks." *IEEE/ACM Transactions on Networking*, vol. 12, no. 4, pp. 609–619, 2004.
17. M. Bhardwaj and A.P. Chandrakasan, "Bounding the lifetime of sensor networks via optimal role assignments," *IEEE INFOCOM*, 2002.
18. M. Cardei, M. T. Thai, Y. Li, W. Wu, "Energy-efficient target coverage in wireless sensor networks," *IEEE INFOCOM*, 2005.
19. X. Wu, G. Chen and S. Das, "Avoiding energy holes in wireless sensor networks with nonuniform node distribution," *IEEE Transactions on Parallel and Distributed Systems*, vol.19, no.5, pp. 710–720, 2008.
20. M. Ma and Y. Yang, "SenCar: An energy efficient data gathering mechanism for large scale multihop sensor networks," *IEEE Transactions on Parallel and Distributed Systems*, vol. 18, no. 10, pp. 1476–1488, October 2007.
21. J. Luo, J. P. Hubaux, "Joint sink mobility and routing to maximize the lifetime of wireless sensor networks: the case of constrained mobility," *IEEE/ACM Transactions on Networking*, vol. 18, no. 3, pp. 871–884, June 2010.
22. M. Zhao and Y. Yang, "Bounded relay hop mobile data gathering in wireless sensor networks," *IEEE Transactions on Computers*, vol. 61, no. 2, pp. 265–277, Feb. 2012.
23. M. Zhao. M. Ma and Y. Yang, "Efficient data gathering with mobile collectors and space-division multiple access technique in wireless sensor networks," *IEEE Transactions on Computers*, vol. 60, no. 3, pp. 400–417, March 2011.
24. M. Zhao and Y. Yang, "Optimization based distributed algorithms for mobile data gathering in wireless sensor networks," *IEEE Transactions on Mobile Computing*, vol. 11, no. 10, pp. 1464–1477, October 2012.
25. M. Ma, Y. Yang and M. Zhao, "Tour planning for mobile data gathering mechanisms in wireless sensor networks," *IEEE Transactions on Vehicular Technology*, vol. 62, no. 4, pp. 1472–1483, May 2013.

26. M. Zhao, Y. Yang and C. Wang, "Mobile data gathering with load balanced clustering and dual data uploading in wireless sensor networks" to appear in *IEEE Transactions on Mobile Computing*, 2015.

27. N. Tesla, "Apparatus for transmitting electrical energy," U.S. Patent 11119732, Dec. 1914.

28. A. Kurs, A. Karalis, R. Moffatt, J. D. Joannopoulos, P. Fisher and M. Soljacic, "Wireless power transfer via strongly coupled magnetic resonances," *Science*, vol. 317, pp. 83, 2007.

29. A. Kurs, R. Moffatt and M. Soljacic, "Simultaneous mid-range power transfer to multiple devices,"*Applied Physics Letter*, vol. 96, no. 4, article 4102, Jan. 2010.

30. Powermat©, "http://www.powermat.com."

31. Powercast Corp©, "http://www.powercastco.com".

32. Hevo power©, "http://www.hevopower.com."

33. B. Tong, Z. Li, G. Wang and W. Zhang, "How wireless power charging technology affects sensor network deployment and routing," *IEEE Distributed Computing Systems (ICDCS)*, 2010.

34. S. He, J. Chen, F. Jiang, D. Yau, G. Xing and Y. Sun,"Energy provisioning in wireless rechargeable sensor networks," *IEEE Transactions on Mobile Computing*, vol. 12, no. 10, pp. 1931–1942, Oct. 2013.

35. L. Fu, P. Cheng, Y. Gu, J. Chen and T. He, "Minimizing charging delay in wireless rechargeable sensor networks," *IEEE INFOCOM*, pp. 2922–2930, 2013.

36. Y. Peng, Z. Li, W. Zhang and D. Qiao, "Prolonging sensor network lifetime through wireless charging," *IEEE Real-Time Systems Symposium (RTSS)*, pp. 129–139, 2010.

37. Z. Li, P. Yang, W. Zhang, and D. Qiao, "J-RoC: a Joint Routing and Charging Scheme to Prolong Sensor Network Lifetime", *IEEE International Conference on Network Protocols (ICNP)*, 2011.

38. H. Dai, Y. Liu, G. Chen, X. Wu and T. He,"Safe charging for wireless power transfer," *IEEE INFOCOM*, pp. 1105–1113, 2014.

39. Y. Shu, P. Cheng, Y. Gu, J. Chen and T. He,"TOC: Localizing wireless rechargeable sensors with time of charge," *IEEE INFOCOM*, pp. 388–396, 2014.

40. Online: "http://www.afar.net/tutorials/fcc-rules".

41. M. Zhao, J. Li and Y. Yang, "Joint mobile energy replenishment and data gathering in wireless rechargeable sensor networks," *IEEE Transactions on Mobile Computing*, vol. 13, no. 12, pp. 2689–2705, 2014.

42. Y. Shi, L. Xie, T. Hou and H. Sherali, "On renewable sensor networks with wireless energy transfer," *IEEE INFOCOM*, pp. 1350–1358, 2011.

43. L. Xie, Y. Shi, T. Hou, W. Lou, H. Sherali and S. Midkiff, "On the renewable sensor networks with wireless energy transfer: the multi-node case," *IEEE International Conference on Sensing, Communication, and Networking (SECON)*, 2012.

44. S. Guo, C. Wang and Y. Yang, "Joint mobile data gathering and energy provisioning in wireless rechargeable sensor networks," *IEEE Transactions on Mobile Computing*, vol. 13, no. 12, pp. 2836–2852, 2014.

45. C. Wang, J. Li, F. Ye and Y. Yang, "NETWRAP: An NDN based real-time wireless recharging framework for wireless sensor networks," *IEEE Transactions on Mobile Computing*, vol. 13, no. 6, pp. 1283–1297, 2014.

Chapter 2
Network Architecture and Principles

2.1 Network Components

Assume that sensor nodes are uniformly and randomly distributed in the network, and nodes are stationary and each node knows its deployed location. For scalable performance, the network is divided into several *areas* and each area is further divided to generate some new *sub-areas*. A new level is generated in each division. The divisions are based on geographical coordinates of the sensing field. An example of a 2-level WRSN network is shown in Fig. 2.1. The two areas represented by solid lines are generated at the first level. Then each area is further split into two sub-areas represented by dashed lines on the second level. Several key network components are explained below.

- *Charging Vehicles*: A charging vehicle has positioning systems (GPS) and knows its location. The sensor locations are pre-processed during network initialization and known to the charging vehicles. The vehicles are equipped with high density battery packs and charging coils. They also have communication capability by launching powerful antennas. In this way, they can not only query the network for node status information but also communicate among themselves or to the base station via long range communication technologies (e.g., cellular, WiMax).
- *Base Station*: The base station is used for collecting sensing data and performing network management. The charging vehicles can be commanded remotely by the network administrator via the base station. It also has computing capabilities to perform the tasks of calculating recharge sequences and dispatching charging vehicles. When a charging vehicle almost depletes its own energy, it returns to the base station for a quick battery replacement.
- *Head Nodes*: A head node is a sensor node that aggregates node's status information in its subordinate area. When requested by a charging vehicle or the head node of its superior level, it aggregates node status information from the subordinate sub-areas at the lower levels and sends to the requester.

© The Author(s) 2015
Y. Yang, C. Wang, *Wireless Rechargeable Sensor Networks*, SpringerBriefs
in Electrical and Computer Engineering, DOI 10.1007/978-3-319-17656-7_2

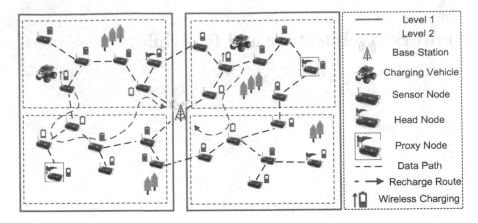

Fig. 2.1 Network architecture

- *Proxy Nodes*: An emergency occurs when a node's battery energy falls below a threshold (e.g., 10 %). It needs to be handled by the charging vehicles immediately. The head nodes on the top-level are selected as proxies so they can aggregate emergency information from sensor nodes directly without propagating through the network hierarchy.
- *Normal Nodes*: A sensor node not selected as a head is a normal node. It reports its status information to its superior head node, or sends emergency information directly to its proxy when the battery energy drops below the recharge threshold.

Let N denote the total number of sensor nodes in the network and L denote the side length of the square sensing field. Then the node density is $\rho = \frac{N}{L^2}$. For event-driven sensing applications with events occurring at each location with equal probability, spatially and temporally independent of each other, the data generation process can be modeled as a Poisson process with average rate λ [1]. All sensors transmit at the same power level with fixed transmission range r. The energy consumed for transmitting/receiving a packet is e_t and e_r, respectively. To obtain their values with respect to packet length l, we can utilize the model in [2]. The base station is placed at the center of the field to collect sensed data in multi-hops. When receiving a status request, a sensor node transmits its status information including energy level and lifetime to the head node in the sub-area. If its energy drops below the emergency threshold, it sends out an emergency recharge request to the proxy node.

The network has m charging vehicles. Once the recharging voltage at the sensor's output circuit is enough to provide a charge, the recharge time is governed by battery characteristics. The typical recharge time required to bring battery energy from zero to full capacity C_s is T_r time (e.g., for a Panasonic Ni-MH AAA battery [3] of battery capacity $C_s = 780$ mAh, $T_r = 78$ min). All charging vehicles are equipped with high-density batteries of $C_h(C_h \gg C_s)$ and consume at e_c J/m while moving at speed v m/s.

2.2 Principles in Wireless Rechargeable Sensor Networks

In this section, we introduce several principles in WRSNs from the theoretical aspects. These include energy neutrality, number of charging vehicles, node lifetime and adaptive recharge threshold.

2.2.1 Energy Neutrality

For a WRSN, the principle of energy neutrality must hold. That is,

$$E(T) \leq R(T) + E_0 \tag{2.1}$$

where T is the time duration, $E(T)$ is the total energy consumption of the network in T, $R(T)$ is the total energy replenished into the network by the charging vehicles in T and E_0 is the initial energy of all the nodes. In other words, the energy neutral condition states that the energy consumption of all the sensor nodes must be less than or equal to the total energy available in a long time perspective. Otherwise, nodes in the network would deplete energy eventually.

We can obtain the number of charging vehicles needed to satisfy Eq. (2.1). First, let us estimate $R(T)$. Since recharge time depends on the specific battery characteristics, the maximum energy a charging vehicle can put back into the network in T_r time is at most C_s. The maximum charging capacity occurs when the vehicle recharges nodes one after another without any idling time in between. The average moving time between two consecutive sensor locations can be estimated through the average distance between two random locations in the square field of length L. From [5], we obtain the average distance $\bar{d} = \frac{2\sqrt{2}+10\ln(\sqrt{2}+1)+4}{30}L \approx 0.52L$. For charging vehicles moving at constant speed v m/minute, the amount of energy replenished into the network is,

$$R(T) = \frac{mC_sT}{0.52L/v + T_r}. \tag{2.2}$$

$E(T)$ on the left hand side of Eq. (2.1) is a random variable since the packet generation process is Poisson. The network with length L has at most $h = \lceil \frac{\sqrt{2}L}{2r} \rceil$ hops to the boundaries. As studied in [4], it can be closely approximated by h concentric rings and each inner ring carries traffic from all outer rings. Since nodes are uniformly and randomly distributed, the number of nodes in the i-th corona, is $N_i = (2i - 1)r^2\pi\rho$ for $0 < i \leq h$. We start with the estimation of energy consumption in each ring. The average energy consumption for the i-th ring is $(0 < i \leq h)$,

$$\mu_i = N_i \lambda T e_t + \sum_{j=i+1}^{h} N_j \lambda T (e_t + e_r)$$

$$= r^2 \pi \rho \lambda T \left[(h^2 - i^2)(e_t + e_r) + (2i - 1)e_t \right] \qquad (2.3)$$

By summing Eq. (2.3) from 1 to h, we obtain the total network energy consumption

$$\overline{E(T)} = \left[\sum_{i=1}^{h} (h^2 - i^2)(e_t + e_r) + (2i - 1)e_r \right] r^2 \pi \rho \lambda T$$

$$= \left[\left(\frac{2}{3}h^3 - \frac{1}{2}h^2 - \frac{1}{6}h \right)(e_t + e_r) + h^2 e_t \right] r^2 \pi \rho \lambda T \qquad (2.4)$$

Note that the derivation of total energy consumption is based on the fact that nodes generate packets independently and randomly following a Poisson process and the sum of Poisson random variables are still Poisson with mean equal to the sum of their average rates.

We can now plug $R(T)$ and $E(T)$ into Eq. (2.1). We have the following theorem.

Theorem 1. *The probability for the energy neutral condition to hold is*

$$P_{op} = \Phi \left(\frac{R(T) + E_0 - \overline{E(T)}}{\sqrt{\overline{E(T)}}} \right) \qquad (2.5)$$

where $R(T)$ and $\overline{E(T)}$ are obtained in Eqs. (2.2) and (2.4), respectively. $\Phi(\cdot)$ denotes the Cumulative Distribution Function of the Normal distribution.

Proof. Energy consumption in the network is taken by the sum of independent Poisson variables over T. When T is observed over a long time period, we can use the *Central Limit Theorem* to approximate Poisson distribution by a Normal distribution $\mathcal{N}(\overline{E(T)}, \overline{E(T)})$ (note that the mean and variance of a Poisson distribution is the same) [6].

From *Theorem 1*, we immediately have the following Proposition.

Proposition 1. *The minimum number of charging vehicles required to maintain perpetual operation is*

$$m = \left\lceil \frac{(\Phi^{-1}(\epsilon)\sqrt{\overline{E(T)}} + \overline{E(T)} - E_0)(0.52L/v + T_r)}{C_s T} \right\rceil \qquad (2.6)$$

where $\Phi^{-1}(\epsilon)$ is the inverse Cumulative Distribution Function of Normal distribution and ϵ is a value very close to 1.

Proof. Since $\Phi^{-1}(1) \to \infty$, we consider the network achieves perpetual operation with a very high probability approaching 1 but not equal to 1, e.g., $\epsilon = 0.99$, $\Phi^{-1}(0.99) \approx 2.33$. From Eq. (2.5), we have

$$\frac{\frac{mC_sT}{0.52L/v+T_r} + E_0 - \overline{E(T)}}{\sqrt{E(T)}} \geq \Phi^{-1}(\epsilon).$$

After some manipulations, the minimum number of charging vehicles, m, needed to satisfy the energy neutral condition can be obtained. This result can be used to calculate the number charging vehicles needed in the network planning stage.

2.2.2 Estimation of Node Lifetime

In this subsection, we introduce a method to estimate how long a sensor node can survive given its current energy level. This information can help us construct effective recharge schedules. Since a node's energy consumption rate is a random variable and depends on traffic patterns, it is important for each node to know its traffic amount which is generally determined by the number of hops from the base station. This information can be obtained by message propagation from the base station using a typical routing protocol and adjusted accordingly during the network operation.

From Eq. (2.3), the average traffic rate of a node in the j-th ring ($1 \leq j \leq h$) can be easily calculated: $\lambda_j = \lambda(1 + (h^2 - j^2)/(2j - 1))$. Given current battery energy E, the maximum number of packets the node can transmit is $n = \lfloor \frac{E}{(e_t + e_r)} \rfloor$.

Theorem 2. *Given a node with energy E at the j-th ring waiting to be recharged, it will survive time t with probability*

$$P(L_j > t) = 1 - \frac{\gamma(n, \lambda_j t)}{\Gamma(n)}, \tag{2.7}$$

where $n = \lfloor \frac{E}{(e_t + e_r)} \rfloor$, $\gamma(\cdot, \cdot)$ and $\Gamma(\cdot)$ are the respective lower incomplete gamma function and complete gamma function [6].

Proof. The summation of interarrival times of packets until the sensor node can no longer transmit packets is the lifetime of the sensor node. Since the data generation process is Poisson with rate λ_j, the interarrival time of packets is exponentially distributed. It is known that the sum of independently identically distributed exponential variables results in a *Gamma distribution* with probability density function

$$f_{L_j}(x) = \lambda_j e^{-\lambda_j x} \frac{(\lambda_j x)^{n-1}}{(n-1)!}, x \geq 0 \tag{2.8}$$

and the Cumulative Distribution Function of Gamma distribution is

$$P(x < t) = \int_0^t \lambda_j e^{-\lambda_j x} \frac{(\lambda_j x)^{n-1}}{(n-1)!} dx = \frac{\gamma(n, \lambda_j t)}{\Gamma(n)} \tag{2.9}$$

Proposition 2. *Let T_l denote the estimated lifetime of a node in a WSRN. For a recharge sequence of N nodes, if a node at the j-th ring has probability $\frac{\gamma(n, \lambda_j T_l)}{\Gamma(n)} \approx 0$, $T_l = (N-1)(T_r + \sqrt{2}L/v)$, no matter where the node is placed in the recharge sequence, it will not deplete battery energy before its recharging starts.*

Proof. The worst case occurs when the node is placed at the end of the recharge sequence. The longest waiting time to get recharged is $T_l = (N-1)(T_r + \sqrt{2}L/v)$ since there are $N-1$ nodes ahead with $\sqrt{2}L/v$ maximum traveling time between two sensor nodes and $\sqrt{2}L$ is the diagonal of the square field. Once $\frac{\gamma(n, \lambda_j T_l)}{\Gamma(n)} \approx 0$, $P(L_j > T_l)$ approaches 1 so it is guaranteed to recharge the node before it depletes battery energy.

Based on *Proposition 2*, given a recharge sequence, the probability that a node can survive the entire recharging process can be calculated. The recharge scheduling algorithm in the following chapters takes this result as an input.

2.2.3 Adaptive Recharge Threshold

In this subsection, we consider the case that nodes with different traffic amount have different recharge thresholds. The difference of energy consumption between nodes at different locations is caused by different traffic load. That is, a node lies in the inner rings closer to the base station would relay more packets, so it is reasonable to have a higher recharge threshold than the nodes in the outer rings. On the other hand, if all the nodes follow a universal recharge threshold, nodes close to the base station would deplete energy very fast and request recharge more often. This would also make the charging vehicles frequently visit these nodes and lead to unnecessary moving. To this end, the recharge thresholds should be set proportionally (adaptively) to energy consumption rates.

Let τ_i $(0 < \tau_j < 1)$ denote the recharge thresholds for nodes at the j-th ring. We let the ratio of recharge thresholds of ring i and ring j equal the ratio of their energy consumption for data transmission. Suppose the recharge threshold of the first ring is τ_1. Then the thresholds for other rings are

$$\tau_i = \frac{(h^2 - i^2)(e_t + e_r) + e_t(2i - 1)}{(h^2 - 1)(e_t + e_r) + e_t}\tau_1 \approx \frac{2h^2 - (i - 1)^2 - i^2}{2h^2 - 1}, \tag{2.10}$$

where $0 < i \leq h$. The approximation is taken under the assumption that $e_t \approx e_r$. To illustrate Eq. (2.10), e.g., $h = 5$, after τ_1 is set, we obtain $\tau_2 = \frac{45}{49}\tau_1$, $\tau_3 = \frac{37}{49}\tau_1$, $\tau_4 = \frac{25}{49}\tau_1$ and $\tau_5 = \frac{9}{49}\tau_1$.

2.3 Summary

We have described basic network components and network model in this chapter. Several important theoretical aspects in Wireless Rechargeable Sensor Networks have been discussed. These include the energy neutral conditions, number of charging vehicles to maintain perpetual operation, estimation of node lifetime and adaptive recharge thresholds. The theoretical results and analysis will be used in the designs of recharge scheduling algorithms later in this book.

References

1. V. Rai and R. N. Mahapartra, "Lifetime modeling of a sensor network,"*Proceedings of IEEE Design, Automation and Test in Europe (DATE)*, vol. 1, 2005.
2. W. R. Heinzelman, A. Chandrakasan and H. Balakrishnan, "Energy-efficient communication protocol for wireless microsensor networks," *IEEE Proceedings of the 33rd Annual Hawaii International Conference on System Sciences (HICSS)*, 2000.
3. Panasonic Ni-MH battery handbook,"http://www2.renovaar.ee/userfiles/Panasonic_Ni-MH_ Handbook.pdf".
4. X. Wu, G. Chen and S. Das, "Avoiding energy holes in wireless sensor networks with nonuniform node distribution," *IEEE Transaction on Parallel and Distributed Systems*, vol.19, no.5, 2008.
5. S. Dunbar, "The average distance between points in geometric figures," *The College Mathematics Journal*, vol. 28, no. 3, 1997, pp. 187–197.
6. S. Ross, *A First Course in Probability*, 8th Ed, Prentice Hall, 2009.

Chapter 3
Distributed Node Status Reporting Protocol

3.1 Overview

To perform effective recharge and maintain network operations, charging vehicles should obtain global node status information of sensors. This information includes residual battery energy, node lifetime, identification, location, etc. Since sensors do not keep track of charging vehicle's locations during operations, a trivial way is to flood the network with status packets periodically. However, for a network with N nodes, $\mathcal{O}(N^3)$ packet transmissions might be needed in the worst case. This is because that the number of edges in a completely connected graph is $\frac{N(N-1)}{2}$ and there are N status packets from different nodes on all the edges. Apparently, the cost becomes prohibitive for any network contains more than a few hundreds of nodes. Indeed, for each recharge maneuver, the charging vehicle only picks a small subset of nodes with immediate energy demands for recharge, status information from other regions could be regarded as useless. If the useless information can be filtered out before reported to the charging vehicles, a great amount of communication overhead can be avoided. Therefore, we introduce a real-time communication protocol for node status gathering in the network.

The charging vehicles obtain the real-time node status information before making any recharge decisions. Node status information is aggregated on *head* nodes at different levels. For robustness, the head node is usually elected with the maximum battery energy in its subordinate area. The head election process is initiated in the network startup phase through propagation of *head election* packets. During the operation, when a head node is low on energy, it will appoint another node with high energy in its area, and send out a *head notification* packet to notify the new head node. The details will be discussed in the next subsection.

To start the information gathering process, charging vehicles send out *status request* packets to poll the *head* nodes on the top-level first. Once the head nodes receive such packets, they generate new *status request* packets for the lower level

© The Author(s) 2015
Y. Yang, C. Wang, *Wireless Rechargeable Sensor Networks*, SpringerBriefs
in Electrical and Computer Engineering, DOI 10.1007/978-3-319-17656-7_3

head nodes in respective subordinate areas. This process repeats down the network hierarchy until the bottom-level *status request* packets reach all the nodes in the bottom-level subareas.

Once a sensor node receives a bottom-level *status request*, it responds by sending out a *status* packet that contains its current energy level, estimated lifetime, identification and position, etc. When the bottom-level head nodes receive such *status* packets, they select sensor nodes with energy level below their corresponding recharge thresholds, and forward their status information in a combined *status* packet to their superior head nodes. This process repeats from the bottom up along the hierarchy until the top-level head nodes successfully aggregate all the status information from designated areas. This information is then sent to the requested charging vehicle. In the case that there are more than one charging vehicles send out such request simultaneously, the top-level head nodes send the aggregated node status information to the vehicle with fewer communication hops. For overhead reduction, the head nodes take partial responsibilities to pre-select nodes for recharge. On the bottom level, the head nodes only report those nodes with energy level below the threshold.

Once a node's energy falls below an emergency threshold (e.g., 10 % of full capacity), without waiting for the charging vehicles to send out request, it preemptively transmits an *emergency* packet to the proxy node that manages its area. The route from each node to its proxy is established by *head election* messages from the proxy and updated during the operation accordingly. Once a charging vehicle finishes recharging a node, it sends out an *emergency request* packet to see whether there is emergency. These packets are directed to the proxy nodes where updated emergency lists are stored and they respond by sending back identifications, lifetime estimations and energy levels to the charging vehicle. The charging vehicle receives this packet and adopts an appropriate recharge scheduling algorithm to decide the recharge sequence.

The mechanism in the head election protocol shares some similarities with [1, 2]. In the following, we describe the new protocols for communication between head nodes on different levels.

3.2 Protocol Design

We describe the protocol design in this section for a network with l levels.

3.2.1 Head Election

At the initialization phase, the network performs head election starting from the bottom l-th level and this process is propagated up to the top level. Each node generates a random number x and compares it with a pre-determined threshold K.

If $x > K$, it floods a *head election* packet in its subarea at the l-th level. The packet contains the random number x and its identification. Then the node sets it as its maximum random number at its local record $x_{max} = x$. Otherwise, if $x \leq K$, the node waits for receiving packets from other nodes.

Upon receiving a *head election* packet, a node first compares the random number field in the packet with its local record x_{max}. If its local record is larger, the packet is discarded. Otherwise, the sensor updates x_{max} to that in the packet accordingly and records the identifier in the packet. Then it sends out the packet to all its neighbors except the one where packet is received from. This process can be regarded as a distributed fashion to elect the node with the maximum x in each subarea on the bottom level.

On the $(l - 1)$-th level, the newly elected head nodes compete for the heads on this level following a similar manner. They flood new *head election* packets in their subareas on the $(l - 1)$-th level. Nodes follow the same procedure to compare the received random number x and finally the head nodes are elected. This process is repeated until the heads on all the levels are elected.

To build intermediate routing information from each node to its head, the head election packets that do not succeed in the comparison are not discarded except for the bottom level. Instead, they are propagated throughout the respective subarea. This ensures the intermediate nodes to know the routes to the head nodes. Once an upper level head node wants to communicate with its subordinate head nodes, these entries in the routing tables on each intermediate node can be utilized.

3.2.2 Status Request

The hierarchical head structure is constructed to facilitate the propagation of *status request* packets. These packets collect the current status from nodes to offer charging vehicles a global view of the network. The status information is gathered on demand. That is, it can be either sent out after a charging vehicle finishes recharging every node or once in a while to reduce communication overhead in the network.

After the head hierarchy is constructed, the charging vehicles send *status request* packets to query nodes that need recharge. Upon receiving such packets, intermediate nodes use the routing tables established during head election process to forward the packets to all top-level head nodes. At the same time, an intermediate node also leaves an entry in its routing table pointing to the neighbor from which the status request packet is received. This entry is used to guide *status* packets back to the charging vehicles. In Fig. 3.1, the propagation status request of a network with two levels is illustrated. After a status request is sent by a charging vehicle, status information is converged from the bottom level to the top level and finally delivered to the charging vehicle.

After receiving a *status request* packet, a top-level head generates a new status request packet and transmits it to its child-heads. These packets use the routing entries set up during the head election process to find the lower-level head nodes.

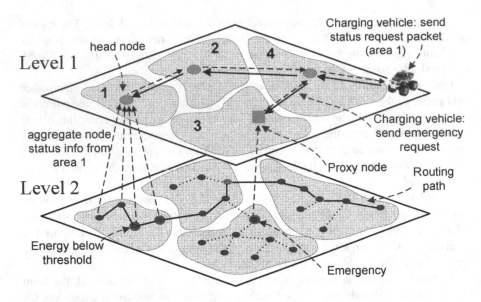

Fig. 3.1 Illustrating propagation of different types of packets

Similarly, nodes also set up routing entries where these packets are coming from so that later status packets can be aggregated at the upper level heads. This process repeats down the head hierarchy until the bottom level heads are reached. Those heads then flood the *status request* packets in their respective subareas.

It could be the case that two or more charging vehicles are requesting node status simultaneously. To avoid receiving duplicated information, we direct the status packets towards the charging vehicle with fewer hop counts. The *status request* packet carries a field to count the hops from the charging vehicle, i.e., the field grows by one at each intermediate node. Once multiple *status request* packets are received by a head node, an intermediate node updates its routing entries by recording only the neighbor with the smallest hop count. In this way, status information from a head node follows the route to reach the charging vehicle with the smallest hop count. Since the charging vehicles are moving during the operation, these routing entries are updated for each status request.

3.2.3   Status Report and Recharge

Once a node receives a bottom level *status request* packet, it responds with information including its current energy level, estimated lifetime, identification and position. These packets are easily routed back to the bottom level heads based on the routing entries set up earlier. The head nodes quickly check if the reported energy level of a node is less than the node's recharge threshold. If so, the identification of the node is added to a local recharge list at the head node, and the energy demand is also added to a cumulative summation counter. Once the head finishes collecting status packets in its subarea, it sends out an aggregated *status* packet to its upper level head node. The aggregated *status* packet contains the information from nodes with energy below their recharge thresholds. Note that a method to compute recharge threshold adaptively is introduced in Sect. 2.2.3.

Upon receiving aggregated status packets from the lower level head node, a head node always selects the one with the largest cumulative energy demand and forwards it upwards the head hierarchy. Finally, the charging vehicle close to a head on the top level receives which subarea has the largest energy demand and proceeds to recharge the nodes based on the recharge algorithms discussed later.

By entitling the head node some responsibilities to filter out some sub-areas, communication overhead can be minimized during the process of gathering node status. This is important since node status information is gathered every once in a while, redundant information would not only enlarge the packet length but also increase the computation complexity of recharge schedules.

Figure 3.1 gives a pictorial illustration of a network with two levels. The charging vehicle sends out a status request to poll all the node status information from area 1. The energy request packet is relayed towards the head node in area 1 by nodes in areas 2 and 4. Upon receiving the energy request, the head node in area 1 aggregates node status in its area and reports to the charging vehicle. The packet is routed back following the same route taken by the status request packet.

3.2.4 Emergency Report and Recharge

Emergency occurs when a node's energy falls below the emergency energy threshold. These nodes should be taken care immediately to prevent them from depleting battery energy. Once an emergency is detected, the node immediately sends out an *emergency* packet with its identification and energy level to the proxy node in its area. The proxy nodes are top level head nodes so the emergency packets do not need to propagate through the head hierarchy. The routing information established earlier during the head election process can be used to direct these packets towards the proxy nodes.

The charging vehicle should frequently check whether there is emergency situation by polling the proxy nodes through *emergency request* packets. In principle, to avoid any missing emergency, the charging vehicles should send out such packets after finishing recharging the current node. Once an intermediate node receives an *emergency request* packet, it updates the local routing entries to record where this packet is coming from. This entry is used to route the emergency report packets from the proxy nodes back to the charging vehicles. Since there could be multiple emergency nodes reported while there are also other normal recharge requests, a charging vehicle needs to handle all the emergency situations within a specified time (e.g., the expected time before next emergency occurs). We introduce several recharge scheduling strategies in the next chapter.

Figure 3.1 also shows an example with a node having emergency in area 3. The node immediately reports to the proxy node and the packet is further forwarded to the charging vehicle upon an emergency request.

3.2.5 Head Hierarchy Maintenance

A head node may run out of energy since it usually engages in more activities than other nodes. In this situation, head re-election is needed. In fact, only the head nodes on the bottom levels compete with each other for the head node on an upper level. Since a head node receives status report from all the nodes in its bottom level subarea, it knows the updated node status in its subarea. To reduce overhead, it can easily appoint the node with the highest energy as the *new head node*. A *head notification* packet is then flooded in the bottom level subarea to notify all the nodes of the new head node.

The generation of the new head triggers a new head election process up the head hierarchy. It floods a new *head election* packet in its subarea. Instead of a random number, the packet carries the current energy level of the participating head node. Following the same procedure, nodes in the subarea compare the energy level in the incoming packet and only store the information with the maximum energy. Then the head node with the highest energy level is elected. If this is the same head node, the process stops to avoid unnecessary overhead. Otherwise, the new head triggers a sequence of head election in the upper level and this process repeats until a new head node is elected on the top level.

3.3 Summary

In this chapter, we introduce a distributed communication protocol that can gather node status information in real-time. The protocol uses different types of packets for communication. Initially, the network is divided hierarchically into different levels and a head node in each area is selected for aggregating node status. A charging vehicle first sends out a status request packet to collect node status from designated areas. The packet propagates along the head hierarchy until the bottom level areas are reached. The node status information is gathered at the head nodes and reported all the way up through the hierarchy until the charging vehicle is reached. In case a node is in emergent status, it preemptively sends out an emergency report to the proxy node on the top level. Upon receiving an emergency request packet, the proxy reports all the emergency nodes to the charging vehicle. This structure enables efficient node status collection and ensures scalability.

References

1. W. R. Heinzelman, A. Chandrakasan and H. Balakrishnan, "Energy-efficient communication protocol for wireless microsensor networks," *IEEE Proceedings of the 33rd Annual Hawaii International Conference on System Sciences (HICSS)*, 2000.
2. O. Younis and S. Fahmy, "Distributed clustering in ad-hoc sensor networks: a hybrid, energy-efficient approach", *IEEE Transations on Mobile Computing*, vol. 3, no. 4, 2004.

Chapter 4
Recharge Scheduling

4.1 Emergency Recharge Scheduling Problem

First, we discuss the optimal recharge policy to handle multiple emergencies. According to Sect. 3.2.4, a node that is on the verge to deplete its battery energy will send an emergency recharge request to the proxy node on the top level. In addition, when the charging vehicle is idle, it polls the proxy node to obtain an emergent recharge node list if there is any. Here, we consider the scenario where there are n emergent nodes to be recharged in T_e time. T_e is defined as the average inter-arrival time of emergencies during operations. The value of T_e can be measured and updated iteratively through the operation by charging vehicles. We assume that the sum of their energy demands is much less than the recharging capacity of the vehicle.

Since the charging vehicle may not finish recharging all n nodes within T_e time, our objective is to maximize the amount of energy refilled into the network in T_e. The problem can be formulated as a classic Orienteering Problem (OP) [1]. OP involves a set of points in the field with different rewards to be visited by a player before time expiration. The objective is to maximize the rewards collected before the time expires. To model OP into our problem, the charging vehicle visits sensor nodes for maximizing energy replenishment (reward) within inter-arrival period of emergency T_e. We consider a graph $G = (V, E)$ where vertex V_i represents the emergent sensor locations. The charging vehicle starts from the original location V_0. E is the edges among sensor nodes. The recharging reward r_i of sensor i is defined as the amount of energy replenished from the current energy level to full capacity. The edge cost is defined to be the traveling time t_{ij} between i and j plus the recharge time of node i (denoted as t_i). In order to be consistent with the original OP formulation, we virtually make the charging vehicle return to the starting location after T_e by adding an edge of zero weight, i.e., the traveling time is $t_{i0} = 0$.

© The Author(s) 2015   25
Y. Yang, C. Wang, *Wireless Rechargeable Sensor Networks*, SpringerBriefs
in Electrical and Computer Engineering, DOI 10.1007/978-3-319-17656-7_4

A decision variable x_{ij} for edge e_{ij} is introduced. $x_{ij} = 1$ if the edge E_{ij} is visited, otherwise, it is 0. Variable u_i is defined as the position of vertex i in the recharging path. The emergency recharge scheduling problem is formulated as follows.

$$\textbf{P1}: \quad \max \sum_{i=1}^{n} \sum_{j=1}^{n} r_i x_{ij}, \tag{4.1}$$

Subject to

$$\sum_{i=1}^{n} x_{0i} = \sum_{i=1}^{n} x_{i0} = 1, \tag{4.2}$$

$$\sum_{i=1}^{n} x_{ik} = \sum_{j=1}^{n} x_{kj} \leq 1, \forall k = 1, 2, \ldots, n \tag{4.3}$$

$$\sum_{i=1}^{n} \sum_{j=1}^{n} (t_{ij} + t_i) x_{ij} \leq T_e, \tag{4.4}$$

$$x_{ij} \in \{0, 1\}, \forall i, j = 1, 2, \ldots, n, \tag{4.5}$$

$$1 \leq u_i \leq n, \forall i = 2, 3, \ldots, n, \tag{4.6}$$

$$u_i - u_j + 1 \leq n(1 - x_{ij}), \forall i, j = 2, 3, \ldots, n. \tag{4.7}$$

Constraint (4.2) guarantees that the recharging path starts from starting position 0 and ends at starting position 0. Constraint (4.3) ensures the connectivity of the path and that every node is visited at most once. Constraint (4.4) makes sure that the time threshold T_e is not exceeded. Constraint (4.5) imposes decision variable x_{ij} to be 0–1 valued. Constraints (4.6) and (4.7) eliminate subtours in the planned route. These subtour elimination constraints are formulated according to [2, 9].

If time T_e is set to infinity, OP is reduced to the classic Traveling Salesmen Problem with Profit which is known to be an NP-hard problem [3]. Therefore, adopting heuristic algorithms can achieve a balance between performance and computation complexity. A few algorithms have been proposed in [4–7] and a survey of the problem is available in [1]. Tsiligirides [4] has developed a stochastic Monte Carlo technique to generate a large number of routes and used the divide-and-conquer method to select the best among them. A center-of-gravity heuristic algorithm is proposed in [5]. Another algorithm consisting of five steps is proposed in [6]. Optimal solutions to the OP using the brand-and-cut method is introduced in [7]. However, these algorithms are quite complex in terms of efficiency and computational time. Given energy restrictions in the network and the urgency to resolve the emergent nodes, a fast and efficient algorithm is more desirable in the context of our problem.

Next, we show OP can be approximated into a Knapsack problem [8] in our problem. The Knapsack problem aims to maximize the value of items into a knapsack with limited size. Each item is associated with a known size. In fact, the recharge time of a node i is much more than the traveling time from vehicle's current location k to i (i.e., $t_i \gg t_{ki}$). For example, replenishing a node to full capacity usually takes around an hour, the traveling time only takes a few minutes at most. Therefore, to maximize the amount of energy replenished within T_e, we can focus on the recharge time of node i. Thus, Constraint (4.4) in the original OP formulation can be rewritten as $\sum_{i=1}^{n} t_i y_i \leq T_e$. Here, recharge time t_i corresponds to the item size and recharge reward r_i is the item value in the Knapsack problem respectively. With this reduction, we have a much simpler formulation.

$$\textbf{P2}: \quad \max \sum_{i=1}^{n} r_i y_i, \tag{4.8}$$

Subject to

$$\sum_{i=1}^{n} t_i y_i \leq T_e. \tag{4.9}$$

Although Knapsack problem is known to be NP-complete [8], we can solve it in polynomial time using dynamic programming techniques. Dynamic programming is a strategy to break down a problem into many recurring small subproblems and solve them in a recursive manner. We define a table R with entry $R(i, t)$ to represent the maximum recharging reward attained with total time duration less than t where $1 \leq i \leq n$ and $1 \leq t \leq T_e$. Our goal is to compute every entry in the table towards the maximum value of $R(n, T_e)$. We set all the entries $R(0, t)$ for $1 \leq t \leq T_e$ to zero initially. For all the i and t in the table, if picking a new node for recharge exceeds T_e, the reward remains unchanged $R(i, t) = R(i - 1, t)$; otherwise, $R(i, t) = \max(R(i-1, t), r_i + R(i-1, t - t_i))$. The pseudo-code of the algorithm is shown in Table 4.1. As there are two loops of size n and T_e, the complexity of the algorithm is $\mathcal{O}(nT_e)$, which is much lower than directly implementing those algorithms designed for OP.

Finally, it is important to examine the accuracy of such approximation. To see how accurate this approximation achieves in our problem, we use brute force to calculate the optimal solution to OP thereby providing a baseline for comparison. Due to exponentially increasing combinations of larger datasets, we manage to test several cases for n varies from 3 to 12. We define the accuracy as $1 - \left| \frac{R_k - R_{op}}{R_{op}} \right|$, where R_k is the solution by Knapsack approximation and R_{op} is the optimal solution by brute force. Table 4.2 shows that the accuracy is over 99 % for different T_e.

Table 4.1 Algorithm to approximate orienteering problem

Input: T_e, recharge time t_i, table R with entry $R(i, t)$, $1 \leq i \leq n$ and $1 \leq t \leq T_e$

Output: maximum recharge reward and recharging nodes

Initialize $R(0, t) = 0, 1 \leq t \leq T_e$

For i from 1 to n

  For t from 1 to T_e

    If $t_i \leq t$, $R(i, t) = \max(R(i - 1, t), r_i + R(i - 1, t - t_i))$

    Else $R(i, t) = R(i - 1, t)$

    End If

  End For

End For

Table 4.2 Accuracy of Knapsack approximations to optimal solutions

# Emergencies n	3 (%)	4 (%)	5 (%)	6 (%)	7 (%)	8 (%)	9 (%)	10 (%)	11 (%)	12 (%)
$T_e = 300$ min	100	100	100	100	100	100	100	100	100	100
$T_e = 400$ min	100	100	100	100	100	99.7	99.6	99.9	99.8	99.7
$T_e = 500$ min	100	100	100	100	100	100	100	99.6	100	100

4.2 Normal Recharge Scheduling

Next, we discuss how to schedule multiple charging vehicles for normal battery recharge. In the process of normal recharge, it is also necessary to prevent nodes in the recharge sequence from depleting battery energy. The objective is to minimize the overall moving cost of charging vehicles while maintaining the perpetual network operation and satisfying a few constraints. The first constraint comes from charging vehicle's limited capacity whereas most of the previous works have ignored the moving energy of the vehicle and the limit of its recharge capacity [11, 12]. These simplifications may cause the charging vehicle to deplete energy en route, become stranded and unable to return to the base station. The second constraint is to meet sensors' dynamic battery deadlines. This would require the vehicle to recharge some nodes earlier than others. For example, depending on the size of recharge sequences, some nodes may need *prioritized recharge* to avoid depleting battery energy. How to place these nodes in the recharge sequence to guarantee optimal and feasible solution is an interesting, yet difficult problem. We formalize it into an optimization problem with these constraints and provide two algorithms to tackle the problem.

By using the method introduced in Sect. 2.2, we are able to estimate the number of charging vehicles, m, needed based on energy balance in the network. After each time the node status information is reported to the charging vehicles, a recharge scheduling problem is formed as follows. We denote the set of charging vehicles as $\mathscr{S} = \{1, 2, \ldots, m\}$ and the set of nodes requesting for recharge as $\mathscr{N} = \{1, 2, \ldots, n\}$. Consider a graph $G = (V, E)$, where vertex V_i ($i \in \mathscr{N}$)

is the location of node i requests for recharge, and E is the set of edges. During the operation, the vehicles could have different starting positions. We introduce an virtual vertex V_0^a as the starting position of vehicle a. The weight of each edge E_{ij} is associated with the moving energy cost c_{ij}, which is proportional to the distance between nodes i and j. c_{0i}^a represents the cost from initial position V_0^a of vehicle a to node i. Since different charging vehicles might have different energy during the run, we denote the battery energy of charging vehicle a as C_a ($C_a \leq C_h$). The value of C_a determines the number of nodes it can recharge before it goes back to the base station for its own battery replacement. The energy demand for node i is denoted as d_i (demand equals a node's total battery capacity minus its residual energy). Each sensor node i has lifetime L_i and A_i is the arrival time of a vehicle at node i. We further introduce two decision variables x_{ij}^a for edge E_{ij} and y_{ia} for vertex V_i. The decision variable x_{ij}^a is 1 if an edge is visited by vehicle a, otherwise, it is 0. The decision variable y_{ia} is 1 if and only if node i is served by vehicle a, otherwise, it is 0. u_i is the position of vertex i in the recharge tour. The objective is to minimize the total moving cost of the charging vehicles while guaranteeing that the recharge capacities of charging vehicles are not exceeded and no sensor node depletes battery energy.

$$\textbf{P1}: \quad \max\left(\sum_{a=1}^{m}\sum_{i=1}^{n}\sum_{j=1}^{n} c_{ij}x_{ij}^a + \sum_{a=1}^{m}\sum_{i=1}^{n} c_{0i}^a x_{0i}^a\right) \tag{4.10}$$

Subject to

$$\sum_{j=1}^{n} x_{0j}^a = 1, a \in \mathscr{S}, \tag{4.11}$$

$$\sum_{i=1}^{n} x_{ik} = \sum_{j=1}^{n} x_{kj} = 1, k \in \mathscr{N}, \tag{4.12}$$

$$\sum_{i=1}^{n} d_i y_{ia} + \sum_{i=1}^{n}\sum_{j=1}^{n} c_{ij}x_{ij}^a + \sum_{i=1}^{n} c_{0i}^a x_{0i}^a \leq C_a, a \in \mathscr{S} \tag{4.13}$$

$$\sum_{a=1}^{m} y_{ia} = 1, i \in \mathscr{N}, \tag{4.14}$$

$$A_i \leq L_i, i \in \mathscr{N}, \tag{4.15}$$

$$x_{ij}^a \in \{0,1\}, i, j \in \mathscr{N}, a \in \mathscr{S}, \tag{4.16}$$

$$y_{ia} \in \{0,1\}, i \in \mathscr{N}, a \in \mathscr{S}, \tag{4.17}$$

$$1 \leq u_i \leq n, i \in \mathscr{N}, \tag{4.18}$$

$$u_i - u_j + (n-m)x_{ij} \leq n - m - 1, i, j \in \mathscr{N}, i \neq j. \tag{4.19}$$

In the above formulation, Constraint (4.11) states that the recharge path for each charging vehicle starts at an initial position 0. Constraint (4.12) ensures the connectivity of the path and every vertex is visited at most once. Constraints (4.13) and (4.14) guarantee the vehicle's battery energy is not depleted and each sensor is recharged by only one charging vehicle. Constraint (4.15) guarantees arrival time of a charging vehicle is within each sensor's lifetime. Constraints (4.16) and (4.17) impose x_{ij} and y_{ia} to be 0–1 valued. Constraints (4.18) and (4.19) eliminate the subtour in the planned routes, which is formulated according to [9]. The problem can be reduced to the classic Traveling Salesmen Problem (TSP) with unlimited recharge capacity and unspecified node's battery deadline. Clearly, since TSP is a well known NP-hard problem [8], the recharge scheduling problem is also NP-hard.

A direct solution to the recharge scheduling problem that accounts for both vehicle's capacity and node's deadline is rare in existing literature due to its hardness. Therefore, we first review some literatures that have partially solved the problem. A similar problem to TSP is the Vehicle Routing Problem (VRP) [10]. In VRP, a fleet of vehicles start from the same depot and visit client locations to deliver goods. The difference between VRP and TSP is that the salesmen in TSP are allowed to start from different locations whereas vehicles usually start from the same location. In addition, the number of vehicles could be undetermined in VRP and more vehicles can be added in order to meet the demands from clients. The Capacitated Vehicle Routing Problem (CVRP) is studied in [13–15]. In [13], a method is proposed to decompose the problem into a convex combination of TSP tours and the tours are examined if the capacity constraint is violated. In [14], tree-based CVRP is studied and a 2-approximation algorithm is proposed. In [15], exact solutions of CVRP are explored by a combination of branch-and-cut and Lagrangian relaxation methods. Time constraint is also important in many VRPs. For example, a store may only accept goods delivery from 9:00AM to 5:00PM during regular business hours. How to schedule the fleet of vehicles to make the deliveries within clients' specified time windows is called Vehicle Routing Problem with Time Windows (VRPTW). The problem is studied in [16–19]. In [16], a local search algorithm is proposed to reduce the computation of checking the feasibility of the time constraint. In [17], a theoretical approach of $3 \log n$-approximation algorithm is sought based on established subroutines (where n is the number of nodes). In [18, 19], a relaxed time constraint that allows late arrivals is considered.

Most of these works adopt standard optimization techniques that are effective for datasets with small size and static inputs. Therefore, the optimization can be done offline by computers with strong computing power. In contrast, the wireless sensing environment is statistical in nature. That is, the inputs of energy request would change for each run and the size of such request could be large. Besides, the charging vehicle's energy declines while moving and recharging sensors. The existing solutions cannot handle these dynamic situations. Further, due to limited computing power on the vehicles, it is not cost-effective and efficient to implement algorithms with high complexity. To this end, our objective is to design algorithms that are suitable to the dynamic nature of the recharge scheduling problem.

There are several challenges to solve this complex problem. The first challenge is that the charging vehicles' energy constantly decreases due to moving and recharging sensor nodes. Thus, the recharge route should be built with caution to reflect the vehicle's current energy level and traveling costs to node locations. The second challenge comes from the dynamics of energy consumption due to data transmissions. Some nodes consume energy at higher rates and have shorter lifetime than others. These nodes usually lie on the main routing path and should be taken care of more frequently than others to maintain the operation of the network. The optimal solution to this problem is between achieving conflicting goals. On one hand, to keep all the nodes running, we need to push the charging vehicles to recharge as many nodes as possible. On the other hand, the desire to reduce overall cost needs to minimize the moving distance of charging vehicles. At the same time, the recharge decisions should account for node's lifetime and vehicle's own battery energy as well. We can see that an ideal solution should achieve a good balance between the two objectives without sacrificing either. In the next subsections, we present two such algorithms.

4.2.1 Weighted-Sum Algorithm

First, we present a fast algorithm that leverages the weighted sum of node's lifetime and vehicle's traveling time. Given a charging vehicle's current location at k and two nodes i and j, there are important metrics to affect their orders in the recharge sequence: the traveling time between k to i, j (t_{ki}, t_{kj}), and their lifetime l_i, l_j. If node j is bound to deplete its battery while node i can still last for a while, the vehicle should recharge j first even if j is located further away than i. Therefore, we can see that to maintain perpetual operations, a trade-off has to be made between meeting sensor's battery deadlines and minimizing vehicle's traveling cost. We introduce a weighted sum w_{ij} below

$$w_{ij} = \alpha t_{ij} + (1 - \alpha)l_j. \tag{4.20}$$

For a charging vehicle residing at node i, w_{ij} is used to decide which node j to recharge next. A node with a smaller weighted value is more desirable and should be visited with higher priority. The weight parameter α affects the choice of recharging schedules. When $\alpha = 1$, the algorithm reduces to the nearest neighbor algorithm that the vehicle always recharges the closest node first regardless of battery deadlines; when $\alpha = 0$, it picks the node with the earliest battery deadline first regardless of the traveling time. When the vehicle detects its own battery is about to deplete, it returns to the base station for battery replacement.

Figure 4.1 shows an example of a charging vehicle with three sensor nodes. The lifetime and the traveling time on each edge are shown in the figure. For demonstration purpose, we vary α from 0, 0.5 to 1 and assume the recharge takes 3,600 s (seconds) to finish. At time 0 s, the vehicle calculates the weight for sensor

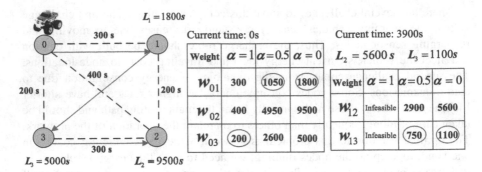

Fig. 4.1 An example of weighted sum algorithm with one charging vehicle and three sensor nodes

nodes 1, 2 and 3. The minimum weights are circled. When $\alpha = 1$, node 3 has the minimum weight of 200 (purely traveling time); when $\alpha = 0.5$, node 1 has the minimum weight of 1,050; when $\alpha = 0$, node 1 also has the minimum weight of 1,800. At this point, if node 3 is visited next, node 1 would have depleted its energy after finishing recharge node 3. Therefore, the choice of $\alpha = 1$ is infeasible in this example and node 1 is visited first. After node 1 has been recharged, choosing node 3 results in the minimum weight for both $\alpha = 0.5$ and $\alpha = 0$. Therefore, the recharge schedule follows $0 - 1 - 3 - 2$ in this example.

We can see that the weight parameter α also affects the feasibility of the solution. Since the total distance is not simply inversely proportional to α, we cannot use a binary search method to locate the best α value. To this end, we find α by searching through a list of candidate α values, A. For example, $\alpha = 0, 0.05, 0.1, 0.15, \ldots, 1.0$, where $|A| = 21$. In this way, a desirable trade-off is achieved between optimality and complexity.

While there are multiple charging vehicles calculating the recharge sequence together, they exchange their location information via long range radio communications. Current technologies such as cellular communications and WiMax can easily realize such coordinations. At the beginning of recharge scheduling, an updated recharge node list is synchronized on all the vehicles. We label the vehicles in orders so they begin the calculation of the next node sequentially. After a vehicle selects a node for recharge, it broadcasts its decision so other vehicles can remove this node in their recharge node list at this point. This operation avoids possible conflicts where multiple charging vehicles select the same node for recharge. Table 4.3 shows the pseudo-code of the entire algorithm.

4.2.2 Adaptive Recharge Scheduling Algorithm

In this subsection, we further provide an adaptive recharge scheduling algorithm. Since energy requests can come from any location in the network, directing the

Table 4.3 Recharge scheduling—weighted sum algorithm

Input: Weight parameter $\alpha \in [0, 1]$ in stepsize $1/(A - 1)$, current position of vehicle at node k, set of energy requests $\mathcal{N}$, traveling time from i to j, t_{ij}, lifetime l_i, $\forall i, j \in \mathcal{M}$.

Output: Recharge sequence Q.

Initialize minDist $= \infty$, obtain updated recharge node list $\mathcal{N}$, set $Q_t = \emptyset$.

For $\alpha = 0, \ldots, 1$ in an increment of $1/(A - 1)$

  While $\mathcal{N} \neq \emptyset$

    $\forall j \in \mathcal{N}$ Compute $w_{kj} \leftarrow \alpha t_{kj} + (1 - \alpha)l_j$.

    Find $j \leftarrow \arg\min_j w_{kj}$. Broadcast node j has been selected to other vehicles.

    Update its local $\mathcal{N} \leftarrow \mathcal{N} - j$, add j to the end of Q_t, move to position j for recharge.

    Update lifetime of the rest nodes $\forall i \in \mathcal{N}$, $l_i \leftarrow l_i - t_{kj} - t_j$.

    If $l_i \leq 0$,

      Declare infeasible and inform base station.

    End If

  End While

  If solution is feasible,

    Compute total cost dist(Q_t).

  End If

  If dist(Q_t) < minDist,

    minDist $\leftarrow$ dist(Q_t), $Q \leftarrow Q_t$.

  End If

End For

If the vehicle's battery is about to deplete, it returns to the base station for battery replacement.

---

charging vehicles to move back and forth in the field for long distance would incur extra moving cost. To this end, we first partition the network adaptively given the energy request. By partitioning the network, the charging vehicles are confined in their own regions so that long distance moving can be avoided. Then to capture vehicle's recharge capacity, we generate Capacitated Minimum Spanning Trees (CMST) on the nodes. The trees preselect which subset of sensor nodes the charging vehicle should recharge to minimize traveling cost and ensure that the total weight of the tree is within the capacity threshold. Finally, we perform route improvements on the nodes in CMST by capturing node's dynamic battery deadlines.

4.2.2.1 Adaptive Network Partitioning

In the first step, based on node status information gathered, the base station can help charging vehicles partition the network into m regions adaptively and assign a working region for each vehicle. The result is then disseminated to vehicles via long range radio. We utilize the well-known *K-means algorithm* to perform the partition [20]. The K-means algorithm is a method to partition the network into different regions with the square sum of distance minimized with respect to the region's centroid. In this way, the vehicle would move in a confined scope with less moving distance.

The objective in the K-means algorithm is to minimize the intra-region square sum of distances between sensor nodes,

$$S = \sum_{j=1}^{m} \sum_{i=1}^{n} \| n_i^{(j)} - \mu^{(j)} \|^2 \tag{4.21}$$

where $\| n_i^{(j)} - \mu^{(j)} \|^2$ is the square distance between a recharge node n_i of region j to the region's centroid $\mu^{(j)}$ (computed by taking the mean of x, y coordinates). The K-means algorithm operates in a recursive fashion.

First, for m charging vehicles, m sensor nodes are selected as the initial centroid of m regions. We can randomly pick m nodes from $\mathcal{N}$. Then we assign each node to the centroid closest to its location. After all the nodes have been assigned to a centroid, we calculate the coordinates of a new centroid again by summing all x and y coordinates and then taking the average. This step is repeated until the centroids no longer change. The resultant centroid is a virtual position that has the minimal sum of distances to all the nodes in the region and the charging vehicle can use it as a starting position to recharge those nodes.

Figure 4.2 shows a snapshot taken during the operation of 70 recharge requests. Four charging vehicles need to cooperate to resolve these energy requests. In the first step, as shown in Fig. 4.3, the network is partitioned into four regions adaptively according to the recharge requests using the K-means algorithm. Each charging vehicle is assigned a region that is close to its current location.

Fig. 4.2 A snapshot of recharge request from sensor nodes

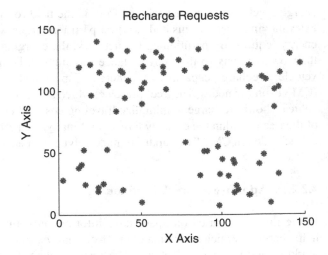

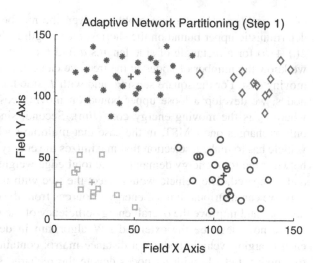

Fig. 4.3 Adaptive network partition according to energy request

4.2.2.2 Generating Capacitated Minimum Spanning Tree

In the second step, after a number of m regions are generated, the vehicles select the closest region from its current location through coordinations. To decide a recharge tour, we need to ensure each charging vehicle's recharge capacity is not exceeded (Eq. (4.13)) and at the same time, we also want to minimize their moving energy cost. To achieve this, we first find the Capacitated Minimum Spanning Tree (CMST) [21] on the energy request. CMST guarantees the sum of energy demands is within charging vehicle's capacity and the minimum moving cost can be found by constructing the minimum spanning tree. In this way, we can ensure sensor nodes close to each other are placed on the same tree and later covered by the same recharge route.

Finding the optimal solution of CMST on a set of nodes is not easy. It requires to search over all possible trees and pick the one with the lowest cost, which involves exponential computations. An efficient algorithm proposed by Esau-Williams (EW) can find a suboptimal solution very close to the optimal solution in polynomial time [21]. The main idea of EW algorithm is to merge any two subtrees when there is a "saving" in the total cost.

The original EW algorithm might have some limitations to apply directly to our problem. First, only the energy demands from nodes are considered while two subtrees are merging whereas the traveling cost on the tree edges is not. Second, multiple CMSTs could be generated for a charging vehicle with limited capacity. Which tree should the charging vehicle select in order to maximize energy efficiency? Here, we introduce an extended EW algorithm in the context of our problem. For the first problem, given the number of nodes in a CMST, a deterministic upper bound on the shortest tour length is derived as $\sqrt{2(n-2)ab} + 2(a+b)$ for a rectangle of side lengths a and b and n nodes in [22]. Thus, once we know the number of nodes in the tree, we can estimate the vehicle's maximum moving cost. For the square sensing field with L side length and a subtree with n_b nodes, we develop a loose upper bound on moving cost, $(\sqrt{2(n_b-2)}+2)Le_c$, where e_c is the moving energy cost (J/m). Second, since each time the vehicle only recharges one CMST, in the case that multiple such trees are generated, the vehicle has to make a selection that maximizes the energy efficiency. Here, the ratio between the total energy demand to the total edge weights of the tree is calculated and compared. The vehicle would select the tree with the maximum ratio. In this way, we can distribute limited energy resources from the charging vehicles into the network and improve the overall energy efficiency of the network.

We now describe the extended EW algorithm in detail. To compute CMST, each charging vehicle updates a distance matrix containing the edge costs of the tree nodes. Let $\mathcal{N}_a$ with n_a nodes denote the recharge set for a charging vehicle ($\bigcup_{a=1}^{m} \mathcal{N}_a = \mathcal{N}$). A trade-off function f_i is defined for each node $f_i = \min(c_{ij}) - c_{0i}$ and $j \in \mathcal{P}_i$. $\mathcal{P}_i$ is the set of neighbors of i. Function $\min(c_{ij})$ finds the minimum cost from node i to its neighbor j in $\mathcal{P}_i$. Function c_{0i} is the cost from node i to the vehicle's starting position (i.e., the root of the tree). To reduce intra-region moving cost, the centroid of the region from network partition is set as the root of the tree. The trade-off function evaluates whether there is a saving of the total cost to merge subtrees of nodes i and j. If $f_i > 0$, it means merging the two subtrees would incur extra cost so it is preferred for the charging vehicle to directly travel from the root to each subtree. If $f_i < 0$, it means merging the two subtrees would have a saving of the total energy cost and the most negative f_i results in the most saving.

Therefore, in each iteration, we compute the trade-off function f_i for each subtree and search through all the values to look for the most negative f_i (i.e., the minimum value). We denote the minimum value by f_k and find node j as k's minimum cost neighbor. If the sum of total demands from the subtrees of k and j plus upper bound of their traveling cost is less than the recharge capacity (which means we can cover the subtrees of k and j under the current recharge capacity), we merge the subtrees of k and j. This step successfully captures the charging vehicle's capacity constraint in Eq. (4.13). Since merging subtrees of k and j results in lower total cost to k, moving from the root directly to k is no longer minimum and should be avoided. The algorithm removes the edge from k to the root by setting the entry c_{0k} in the distance matrix to ∞.

After two subtrees have been successfully merged, the minimum cost from all those tree nodes to the root should be updated. This is achieved by updating the minimum cost in the distance matrix from the tree nodes to the root by setting the value to $\min(c_{0i})$, where i are all nodes in the newly merged tree.

On the other hand, if merging subtrees of k and j causes a violation to exceed vehicle's capacity, any further actions to merge k to j should be restricted because these two subtrees cannot be covered by the charging vehicle in a single run. Then we recompute the trade-off function to look for the next neighboring node that

Table 4.4 Extended Esau-Williams algorithm for charging vehicle a

Input: Recharging node set $\mathcal{N}_a$, distance matrix D, recharge capacity C_a, energy demand $d_i, i \in \mathcal{N}_a$.

Output: CMST with the maximum ratio between energy demands and sum of edge costs.

Initialize trade-off function $f < 0$, weight of each tree, $C_i = 0$.

While (all $f_i < 0$)

  Find neighbor m_i of i results min cost, $\min\limits_{m_i} D(i, m_i)$.

  Compute trade-off value list $f_i = D(i, m_i) - D(1, i), \forall i \in \mathcal{N}_a$.

  Find k and j resulting most negative trade-off value, $k \leftarrow \min\limits_i(f), j \leftarrow m_k$.

  Do

    Add new nodes $N_{new} \leftarrow k + j$ if not exist in current trees

    If weight of merging subtree of $N_{new} < C_a$

    Add N_{new} to tree i, update cumulative weight of i, C_i,

    Declare N_{new} is accepted.

    Else

    Update $D(k, j) \leftarrow \infty$, search for the next min cost neighbor for k.

    $m_k \leftarrow \min\limits_{m_k} D(k, m_k)$, recompute trade-off for k, $f_k = D(k, m_k) - D(1, k)$,

    Declare N_{new} is rejected.

    End If

  Until (N_{new} is accepted) or (all $f_i \geq 0$)

End While

Select a tree results maximum ratio between energy demands and sum of edge costs.

results in minimum trade-off until the next valid neighboring node j is found and merged to the existing trees. The iteration continues until all the trade-offs become nonnegative, in other words, no more saving can be made.

If multiple CMST are generated, the charging vehicle selects a tree with the maximal ratio of energy demand to the sum of tree's edge cost. Later, nodes in this tree are recharged first by forming a recharge route using the route improvement algorithm introduced next. After the charging vehicle finishes recharging all the nodes in a tree, it checks whether its energy falls below a threshold. If so, it returns to the base station for battery replacement. Table 4.4 shows the pseudo-code of the extended EW algorithm.

Figure 4.4 constructs the CMST on the energy request. Note that the CMST is constructed in a parallel fashion so that each vehicle only works on a subset of the total recharging node set.

4.2.2.3 Insertion Algorithm for Recharge Route Improvement

After the CMST is computed, the next step is to find a recharge sequence such that no sensor node would deplete its battery energy during the recharging process. For a CMST, let $\mathcal{N}_c$ denote the node set for charging vehicle a ($\mathcal{N}_c \subseteq \mathcal{N}_a$). From the theoretical principles discussed earlier, we know that if a node's lifetime satisfies

Fig. 4.4 Each vehicle
constructs CMST in its
designated region

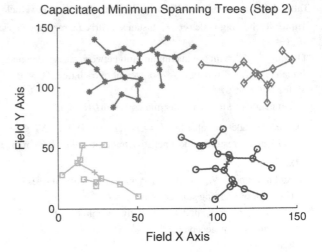

Capacitated Minimum Spanning Trees (Step 2)

conditions *Theorem 2*, the node can be placed anywhere in the recharge sequence
without depleting its energy. These nodes form a *feasible node set* $\mathcal{N}_f$. Otherwise, a
node needs to be prioritized in the recharge sequence according to its lifetime. Note
that nodes report emergency during normal recharging process would most likely
need the prioritized recharge. These nodes form a prioritized set denoted by $\mathcal{N}_p$ and
$\mathcal{N}_f \cup \mathcal{N}_p = \mathcal{N}_c$.

First, since the particular recharging order of feasible nodes does not matter,
we first find the shortest path among them using a TSP algorithm (e.g., $\mathcal{O}(n^2)$
nearest neighbor heuristic algorithm [23], where n is the number of nodes). The
result from the TSP solution serves as the initial sequence from the feasible node
set and the recharge sequence of the shortest path is denoted by Ψ. The next step is
very important. The algorithm needs to insert each node from the prioritized set $\mathcal{N}_p$
into Ψ and makes sure each insertion does not violate the overall time feasibility
imposed by nodes' battery deadlines (Eq. (4.15)). Nodes in $\mathcal{N}_p$ are first sorted in a
descending order regarding their lifetimes and the sorted sequence is denoted as Ω.
Starting from the first node in Ω, nodes are inserted one after another into Ψ. Let
A_i denote the arrival time of the charging vehicle at the i-th node in the shortest
path Ψ, $i = \{1, 2, \ldots, n_f\}$. To insert the j-th node Ω_j from Ω into Ψ, we first
find a location m_t in Ψ such that $A_{m_t} \leq l_{\Omega_j}$ and $A_{m_t+1} > l_{\Omega_j}$ where l_{Ω_j} is Ω_j's
lifetime. We call m_t the *tentative maximum position* to insert Ω_j. It indicates the
maximum number of nodes in Ψ that can be served before node Ω_j depletes its
battery. Since it is possible that all the remaining $|\Omega| - j$ nodes in Ω could be
inserted before Ω_j, the sum of their recharge time would elongate the service time
of Ω_j and might make Ω_j to deplete battery energy prematurely. Therefore, the
maximum position for Ω_j should accommodate the recharge time from all later
insertions. Based on m_t, we further look for the *maximum position* m such that
$A_m \leq A_{m_t} - \sum_{i=j+1}^{n_p} t_i$ and $A_{m+1} > A_{m_t} - \sum_{i=j+1}^{n_p} t_i$, where t_i is the recharge

Table 4.5 Route improvement insertion subalgorithm

Input: CMST $\mathcal{N}_c$, lifetime l_i and recharge time t_i, $i \in \mathcal{N}_c$,
distance matrix D, feasible set $\mathcal{N}_f$ satisfying Proposition 2 in Section 2.2.2.
Output: Recharge sequence Ψ.
Compute the shortest path of the feasible set, $\Psi \leftarrow \text{TSP}(\mathcal{N}_f)$, Sort $\mathcal{N}_p$ in a descending
order regarding lifetime as Ω, initialize $i \leftarrow 1$, node position $k \leftarrow \infty$ in the last step.
While $\Omega \neq \emptyset$
   Find tentative max position m_t in Ψ such that $A_{m_t} \leq l_{\Omega_i}$ and $A_{m_t+1} > l_{\Omega_i}$
   Further find the max insertion position m such that
   $A_m \leq A_{m_t} - \sum_{k=i+1}^{n_p} t_k$ and $A_{m+1} > A_{m_t} - \sum_{k=i+1}^{n_p} t_k$.
   If Cannot find $m \geq 0$.
      Break, return infeasible and report.
   End If
   Set minimum cost $c_{min} \leftarrow \infty$.
   For x from 0 to m
      Insert Ω_i into Ψ, get temporary sequence Ψ_t, calculate cost $c \leftarrow \sum_{j=1}^{|\Psi_t|-1} D(j, j+1)$.
      If $c < c_{min}$, $\Psi \leftarrow \Psi_t, c_{min} \leftarrow c, k \leftarrow x$.   **End If**
   End For
   $i \leftarrow i + 1$, update $\Omega \leftarrow \Omega - i$
End While
Return recharge sequence Ψ, minimum cost c_{min}.

time of Ω_j. At this point, node Ω_j can be inserted into Ψ with the maximum position m. Among these positions, the one with minimum cost is selected as the final insertion position. A new sequence Ψ is obtained after the insertion and Ω_j is removed from Ω. The iteration continues until Ω is exhausted or an infeasible solution is encountered. Table 4.5 shows the pseudo-code of the route improvement insertion algorithm.

Figure 4.5 illustrates how the insertion algorithm works. We consider the prioritized set with two nodes Ω_1 and Ω_2 with lifetime 80 and 65 min, respectively. They need to be inserted into the feasible recharge sequence. We first find that the tentative maximum position k' to insert Ω_1 is between nodes 5 and 6 since $A_5 < l_{\Omega_1} < A_6$. To accommodate the insertion of Ω_2 later, the maximum position k that Ω_1 can be inserted is between nodes 4 and 5 (since $A_3 < A_5 - t_{\Omega_2} < A_4$). Then we search all the four possible locations (before nodes 1, 2, 3 and 4) and find that the position after node 2 and before node 3 minimizes the moving cost. Thus Ω_1 is inserted between nodes 2 and 3. We repeat the procedure for Ω_2. Since it is the only node left in the sorted list, we can directly calculate the maximum position k and find the minimum cost insertion position.

For the CMST in Figs. 4.4 and 4.6 shows the results of improved recharging routes on the selected trees.

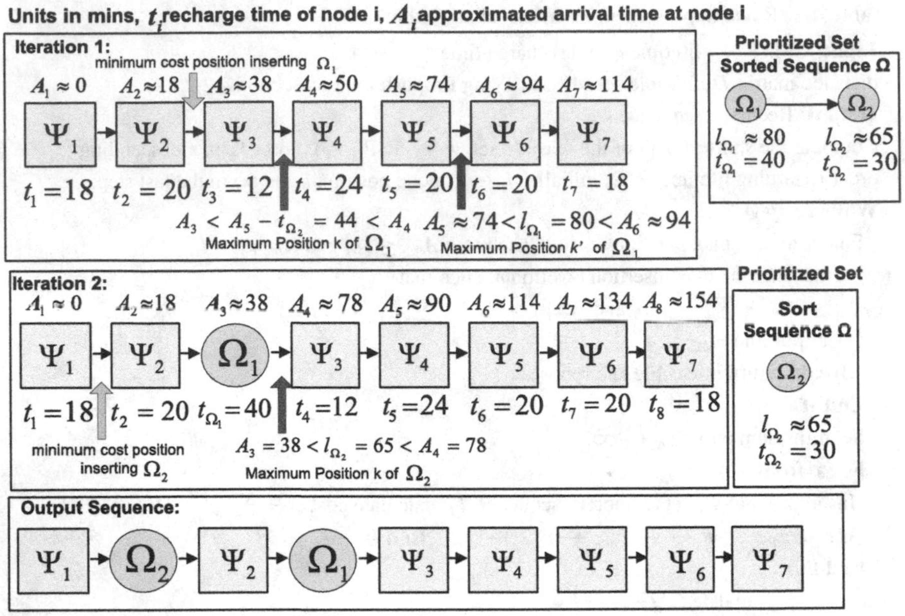

Units in mins, t_i recharge time of node i, A_i approximated arrival time at node i
Iteration 1:
minimum cost position inserting Ω_1
$A_1 \approx 0$ $A_2 \approx 18$ $A_3 \approx 38$ $A_4 \approx 50$ $A_5 \approx 74$ $A_6 \approx 94$ $A_7 \approx 114$
Ψ_1 Ψ_2 Ψ_3 Ψ_4 Ψ_5 Ψ_6 Ψ_7
$t_1 = 18$ $t_2 = 20$ $t_3 = 12$ $t_4 = 24$ $t_5 = 20$ $t_6 = 20$ $t_7 = 18$
$A_3 < A_5 - t_{\Omega_2} = 44 < A_4$ $A_5 \approx 74 < l_{\Omega_1} = 80 < A_6 \approx 94$
Maximum Position k of Ω_1
Maximum Position k' of Ω_1
Prioritized Set
Sorted Sequence Ω
Ω_1 Ω_2
$l_{\Omega_1} \approx 80$ $l_{\Omega_2} \approx 65$
$t_{\Omega_1} = 40$ $t_{\Omega_2} = 30$
Iteration 2:
$A_1 \approx 0$ $A_2 \approx 18$ $A_3 \approx 38$ $A_4 \approx 78$ $A_5 \approx 90$ $A_6 \approx 114$ $A_7 \approx 134$ $A_8 \approx 154$
Ψ_1 Ψ_2 Ω_1 Ψ_3 Ψ_4 Ψ_5 Ψ_6 Ψ_7
$t_1 = 18$ $t_2 = 20$ $t_{\Omega_1} = 40$ $t_4 = 12$ $t_5 = 24$ $t_6 = 20$ $t_7 = 20$ $t_8 = 18$
minimum cost position inserting Ω_2
$A_3 = 38 < l_{\Omega_2} = 65 < A_4 = 78$
Maximum Position k of Ω_2
Prioritized Set
Sort Sequence Ω
Ω_2
$l_{\Omega_2} \approx 65$
$t_{\Omega_2} = 30$
Output Sequence:
Ψ_1 Ω_2 Ψ_2 Ω_1 Ψ_3 Ψ_4 Ψ_5 Ψ_6 Ψ_7

Fig. 4.5 Illustration of insertion algorithm for recharge route improvement

Fig. 4.6 Improve recharge routes by node insertion

Generate Recharging Routes (Step 3)
150
100
50
0
Field Y Axis
0
50
100
150
Field X Axis

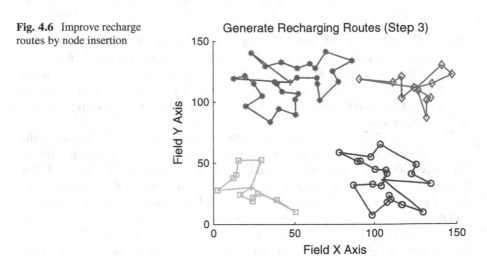

4.2.2.4 Computation Complexity

The time complexity of the two algorithms can be analyzed as follows. The worst case occurs when there is only one charging vehicle available to recharge all N nodes. In the weighted sum algorithm, node selection takes $\mathcal{O}(N^2)$ time and there are a total number of A tests needed. Therefore, the time complexity of the weighted sum algorithm is $\mathcal{O}(AN^2)$.

For the adaptive recharge scheduling algorithm, the extended EW algorithm requires $(N^2 + 2N)$ iterations to find the minimum trade-off value at the outer loop. In the inner loop, the worst case is that for a node with the minimum trade-off value, all its neighbors are rejected due to capacity violations. So N iterations are required. Thus, the extended EW algorithm has time complexity $\mathcal{O}(N^3)$. For the route improvement algorithm, running a TSP algorithm requires $\mathcal{O}(N^2)$ time. Sorting nodes' lifetimes requires $\mathcal{O}(N \log N)$ time and insertion requires $\mathcal{O}(N^2)$ time. Hence, time complexity of the route improvement algorithm is $\mathcal{O}(N^2)$ and the adaptive recharge scheduling algorithm takes $\mathcal{O}(N^3)$ time. When A is less than N, weighted sum algorithm is faster than the adaptive algorithm.

4.3 Summary

In this chapter, we have discussed several recharge scheduling algorithms for different scenarios. In case of emergency recharge, a charging vehicle needs to resolve multiple emergencies at different locations. The problem is formulated into the classic Orienteering Problem that aims to maximize the total amount of recharged energy in a given time period. Based on the fact that recharging time is much larger than traveling time, the problem can be simplified into a Knapsack problem solved by dynamic programming with high accuracy.

In the meanwhile, normal recharge operations requires the charging vehicle to account for its own recharge capacity and different sensor's battery deadlines. It is formulated into a Capacitated Vehicle Routing Problem with Battery Deadlines and two algorithms are provided. The first algorithm leverages a weighted sum of sensor's lifetime and vehicle's traveling time and tries to minimize the weighted value at each recharge. The second algorithm adaptively partitions the network, constructs Capacitated Minimum Spanning Trees and improves the recharge route finally.

References

1. P. Vansteenwegen, W. Souffriau and D. Van Oudheusden, "The orienteering problem: a survey," *European Journal of Operation Research*, vol. 209, no. 1, 2011.
2. C. Miller, A. Tucker and R. Zemlin, "Integer programming formulations and travelling salesman problems," *Journal of the ACM*, pp. 326–329, 1960.

3. D. Feillet, P. Dejax and M. Gendreau, "Traveling salesman problems with profits," *Transportation Science*, vol. 39, no. 2, 2005.
4. T. Tsiligirides, "Heuristic methods applied to orienteering," *Journal of the Operation Research Society*, vol. 35, no. 9, pp. 797–809, 1984.
5. B. Golden, A. Assad and R. Dahl, "The orienteering problem," *Naval Research Logistics* 34, pp. 307–318, 1987.
6. I. Chao, B. Golden and E. Wasil, "A fast and effective heuristic for the orienteering problem," *European Journal of Operation Research*, vol. 88, no. 3, 1996.
7. M. Fischetti, J. S. Gonzalez and P. Toth, "Solving the orienteering problem through branch-and-cut," *INFORMS Journal on Computing*, vol. 10, no. 2, pp. 133–148, 1998.
8. R. Karp. "Reducibility Among Combinatorial Problems," *Complexity of Computer Computations*, pp. 85–103, 1972.
9. B. Gavish, "A note on the formulation of the m-salesman traveling salesman problem," *Management Science,* 1976.
10. P. Toth and V. Daniele, "The vehicle routing problem," *Society for Industrial and Applied Mathematics*, 2001.
11. M. Zhao, J. Li and Y. Yang, "Joint mobile energy replenishment and data gathering in wireless rechargeable sensor networks," *IEEE Transactions on Mobile Computing*, vol. 13, no. 12, pp. 2689–2705, 2014.
12. C. Wang, J. Li, F. Ye and Y. Yang, "NETWRAP: An NDN based real-time wireless recharging framework for wireless sensor networks," *IEEE Transactions on Mobile Computing*, vol. 13, no. 6, pp. 1283–2852, 2014.
13. T. Ralph, "On the capacitated vehicle routing problem," *Mathematical programming*, vol. 94 no. 2–3, pp. 343–359, 2003.
14. B. Chandran and S. Raghavan, "Modeling and solving the capacitated vehicle routing problem on trees," *The Vehicle Routing Problem: Latest Advances and New Challenges*, pp 239–261, 2008.
15. R. Fukasawa, "Robust branch-and-cut-and-price for the capacitated vehicle routing problem," *Mathematical programming*, vol. 106, no. 3, pp. 491–511, May, 2006.
16. M.W.P. Savelsbergh, "Local search in routing problems with time windows," *Annals of Operation Research*, pp. 285–305, 1985.
17. N. Bansal, A. Blum, S. Chawla and A. Meyerson, "Approximation algorithms for Deadline-TSP and Vehicle Routing with Time Windows," *ACM Symposium on Theory of Computing (STOC)*, 2004.
18. H.C. Lau, M. Sim and K. M. Teo,"Vehicle routing problem with time windows and a limited number of vehicles",*European Journal of Operation Research*, 148, pp. 559–569, 2003.
19. E. Taillard, P. Badeau, M. Gendreau, F. Geurtin and J.Y. Potvin, "A tabu search heuristic for the vehicle routing problem with soft time windows,"*Transportation Science*, vol. 31, pp. 170–186, 1997.
20. J. MacQueen, "Some methods for classification and analysis of multivariate observations," *Proceedings of the 5th Berkeley Symposium on Mathematical Statistics and Probability*, vol. 1, no. 14, 1967.
21. L.R. Esau and K.C. Williams, "On teleprocessing system design: part II-a method for approximating the optimal network," *IBM System Journal*, vol. 5, pp. 142–147, 1966.
22. P. Jaillet, "Probabilistic traveling salesman problem," Ph.D. Dissertation, MIT, 1985.
23. T.H. Cormen, C. E. Leiserson, R. L. Rivest and C. Stein, *Introduction to Algorithms*, MIT Press, 2001.

Chapter 5
Performance Evaluations

5.1 Parameter Settings

In the evaluation, 500 sensor nodes are uniformly and randomly deployed over a square sensing field with 200 m side length. The transmission distance is set to 15 m. Nodes use Dijsktra's shortest path algorithm [2] to route data packets to the base station at an average rate of $\lambda = 3$ pkt/min following the Poisson process. Time is equally slotted and each time slot is 1 min. Following the relationship described in Sect. 2.2.3, nodes have adaptive recharge thresholds regarding their hop counts to the base station. Nodes are equipped with CC2430 communication module that draws 27 mA at 3 V while in operative mode [4]. The packets have the same length of 30 bytes and transmission bit rate is 1 Kbps so the energy consumption for transmitting and receiving is 2 mJ.

Once the charging voltage at sensor's reception circuit is enough to provide an effective charge, the recharge time depends on the specific battery characteristics. An off-the-shelf Panasonic AAA battery model is used. The recharge curves can be obtained from [1] with a maximum recharge time at 78 min. Using curve fitting in MATLAB, we obtain a very close approximation of battery recharge time shown in Fig. 5.1.

The charging vehicle carries high density battery packs, e.g., standard 12A, 5 V battery, and consumes at a rate of 5 J/m energy while moving at a constant speed of 1 m/s (calculated in [3]). To see how the performance is affected by the number of charging vehicles, m is varied from 1 to 5 and the simulations are run for 4 months. In the weighted sum algorithm, the weighted parameter α changes from 0 to 1 in an increment of 0.01 so $A = 101$.

© The Author(s) 2015
Y. Yang, C. Wang, *Wireless Rechargeable Sensor Networks*, SpringerBriefs
in Electrical and Computer Engineering, DOI 10.1007/978-3-319-17656-7_5

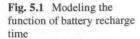

Fig. 5.1 Modeling the
function of battery recharge
time

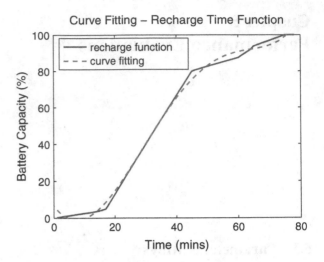

5.2 Comparison of Recharge Scheduling Algorithms

We first examine the performance of the recharge scheduling algorithms in minimizing the moving cost of charging vehicles and compare it with the optimal solution.
Due to NP-hardness of the recharge scheduling problem, acquiring optimal solution
requires exponentially increasing computational efforts with respect to the number
of recharge requests. It becomes prohibitive for more than 10 requests. To this
end, the optimal results for some small-size networks are obtained as baselines
for comparison purpose. The recharge requests are set to emerge uniformly and
randomly from any location in the sensing field and the results are averaged over
100 simulation runs. Figure 5.2 shows the sum of moving energy consumption on
the charging vehicles for the optimal solution and the results from the weighted sum
and adaptive algorithms. The adaptive algorithm provides an average of 1.05 ratio
to the optimal solution whereas the weighted sum algorithm consumes an additional
10 % more energy. Thus, although the adaptive algorithm is a little more complex
than the weighted algorithm, it is able to achieve very close approximation to the
optimal solution.

5.3 Node Nonfunctionality

When a sensor node depletes battery energy, it is not functional anymore until
the next recharge. Nonfunctional nodes may lead to loss of connectivity, network
congestion and disruption so they should be avoided during the operation. Figure 5.3
compares the percentage of nonfunctional nodes between the weighted sum and
adaptive recharge scheduling algorithms. For the weighted sum algorithm, when

Fig. 5.2 Compare
performance of algorithms
with the optimal solution

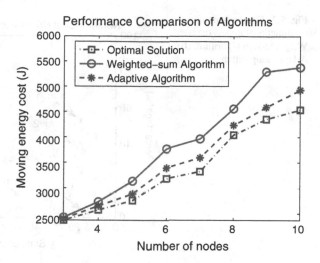

$m = 1$–3, there is a surge of nonfunctional nodes to over 20 % around 20 days until the network stabilizes after 40 days. The spike is due to a majority of sensors request for recharge around the same period of time so the recharge capacities of the charging vehicles are temporarily exceeded. In contrast, the adaptive algorithm provides better performance. For $m = 2, 3$, the spikes disappear and the percentage of nonfunctional nodes is contained within 10 % at network equilibrium. The improvement is because that, although the weighted sum algorithm takes node lifetime as a factor to compute the recharge sequence, it may not select the optimal node for recharge due to the choice of weighted parameter α. That is, the charging vehicles only serve the node with the least weight value each time but this choice only reflects a compromise between traveling cost and lifetime priority, which could be far from optimal for the entire recharge sequence. The adaptive algorithm has taken battery deadline into consideration by inserting nodes that need priority recharge into an established node sequence and guaranteed that each insertion does not violate the time feasibility constraint. It is worth pointing out that when $m = 5$, the adaptive algorithm can maintain the perpetual operation for all the nodes (keeping nonfunctional nodes at zero).

5.4 Energy Evolution

The trace of energy status in the network is demonstrated in Fig. 5.4. Due to the similarity of curve shapes, we have plotted evolution of energy consumption and recharge for the adaptive algorithm in Fig. 5.4a and their cumulative value in Fig. 5.4b. When $m = 1$, it is not sufficient to sustain network operations so nodes deplete energy and become nonfunctional. This corresponds to the drop on the energy consumption curve during the first 10 days for $m = 1$ in Fig. 5.4a. The recharge capacity from only one charging vehicle can barely suffice all the energy

Fig. 5.3 Comparison of nonfunctional nodes. (**a**) Weighted-sum algorithm. (**b**) Adaptive algorithm

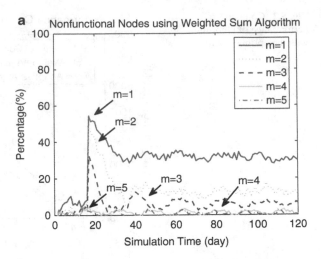

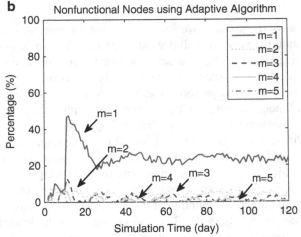

demands so it puts an upper limit on the total energy consumptions in the network and the two curves reach an equilibrium at about 40 days. For $m = 5$, around 5 times energy is recharged into the network and each drop of energy consumption (indicating when there are nonfunctional nodes) corresponds to a surge in energy recharge. This represents $m = 5$ can resolve nonfunctional nodes more effectively.

The cumulative energy status in the network for $m = 1$ and 5 is shown in Fig. 5.4b. To visualize the gaps between curves, the curves in the first 30 days are plotted. If the energy consumption curve is above the recharging curve, more energy has been consumed by the nodes than that has been refilled into the network, and vice versa. For $m = 1$, the energy consumption curve is above the recharge curve. A much wider gap is observed during the first 10 days and as soon as nodes begin to deplete their battery energy, the gap becomes smaller. For $m = 5$, the recharging curve always stays above the energy consumption curve.

Fig. 5.4 Energy evolution in the network when the number of charging vehicle is $m = 1, 5$. (a) Trace of energy evolution. (b) Trace of cumulative energy status (first 30 days)

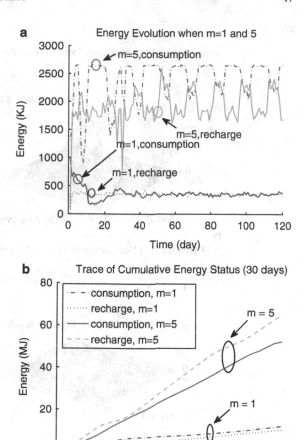

5.5 Duration of Nonfunctional Status

In this section, we consider the percentage of duration that nodes are nonfunctional to the entire simulation time. The results for both algorithms when there are five charging vehicles are plotted in Fig. 5.5 with respect to nodes' locations. First, we can see under adaptive algorithm, nodes have a maximum of 0.6 % time in nonfunctional status. However, the weighted sum algorithm has resulted in a maximum of 3.28 % time in nonfunctional status (5 times more). Further, the shape of the curves indicates that the adaptive algorithm can spread nonfunctional nodes more evenly. A high concentration of nonfunctional nodes near the base station is observed in the weighted sum algorithm. They would easily cause congestions and unavailability of the routing paths to the base station.

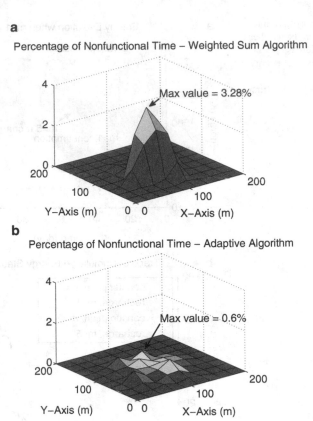

Fig. 5.5 Evaluation of percentage time that nodes are nonfunctional. (**a**) Weighted-sum algorithm. (**b**) Adaptive algorithm

5.6   Data Collection Latency

Data collection latency depends on whether a routing path is available. To successfully transmit all sensed data to the base station timely, all the nodes on the routing paths should be functional. If a node depletes battery energy and no alternate route is available, the packets will be buffered at sensors until the path is restored by the charging vehicle. Shortest path trees are formed using the Dijsktra's algorithm rooted at the base station. Figure 5.6 shows the average data collection delay over the entire simulation time. Since the adaptive algorithm has higher capability of handling nonfunctional nodes than the weighted sum algorithm, 45 % less data delay is achieved with the adaptive algorithm.

Fig. 5.6 Evaluation of data
collection delay

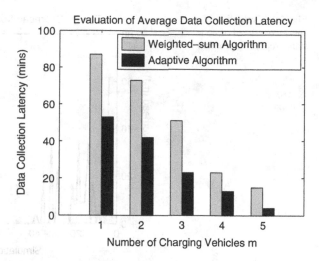

5.7   Overhead of Node Status Collection Protocol

The communication overhead represents the energy consumed for transmitting all types of control packets, e.g., head selection, status request/report, emergency request/report and head notification packets. Figure 5.7 shows the evolution of communication energy overhead during 4 months simulation time. First, compared to the energy cost for transmitting sensed data, the overhead in the communication protocol is not significant. Its energy consumption is around 10 mJ/day compared to at least 8.64 J/day for transmitting sensed data packets. Second, there is a certain amount of overhead during the network setup phase (around 40–60 mJ/day per node) due to the head selection process. During the operation, when a head node is low on energy, it sends out a head notification packet to appoint a new head node. The head selection would then propagate up through the hierarchy. This process also contributes to the overhead from 20 to 120 days time. Further, for different number of charging vehicles, more communication overhead is observed when the number of charging vehicles is not enough ($m = 1, 2$). In this case, recharge requests and head re-selection are performed more often than $m = 3$–4 so higher communication overhead is expected for fewer charging vehicles.

5.8   Charging Vehicle's Moving Energy Cost

We now examine the moving energy cost on the charging vehicles. Since the vehicles also consume energy while moving, one of the goals is to design recharge scheduling algorithms that can provide extra savings in the charging vehicle's moving cost. Figure 5.8 shows the average energy cost per vehicle for the weighted sum and adaptive algorithms. It is interesting to see that when $m = 1$–3, the

Fig. 5.7 Evaluation of
energy overhead of node
status reporting protocol

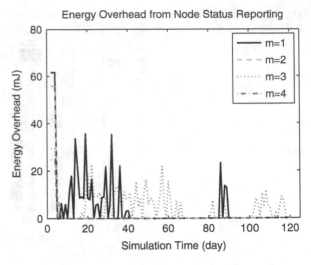

Fig. 5.8 Evaluation of
charging vehicle's moving
cost for weighted sum and
adaptive algorithms

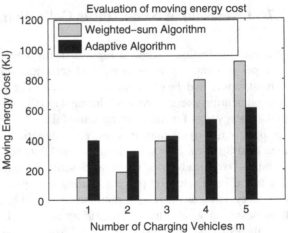

weighted sum algorithm is more energy efficient than the adaptive algorithm. This
is because that when the number of vehicles is not enough ($m = 1$–3) to support
perpetual operation, there are always energy requests so the weighted sum algorithm
selects the nearest node with the least lifetime. When $m = 4$–5, the energy
request becomes sporadic during operations. The weighted sum algorithm that only
recharges the node with the least weighted value might make the vehicle move
long distance. In contrast, the adaptive algorithm partitions the network into smaller
regions so the vehicles have much shorter moving distance in each recharge. This
reduces the moving cost significantly.

5.9 Comparison with Static Optimization Approach

Finally, we compare the performance of the two algorithms with the static optimization approach used in [5]. In the static approach, a charging vehicle selects nodes with energy less than the normal recharge threshold and calculates the shortest recharging tour among the selected nodes. We first compare the percentage of nonfunctional nodes shown in Fig. 5.9. The number of nonfunctional nodes is much higher for the static approach. For $m = 1$, there are 60 % nonfunctional nodes whereas the weighted-sum and adaptive algorithms result in less than 30 % nonfunctional nodes. This is because the static approach does not consider node lifetime and only follows a pre-computed sequence to recharge nodes.

Next, we look at the emergency response time which is the duration from a node reports emergency until it is recharged. Shorter response time means the charging vehicle can respond rapidly to emergencies. Figure 5.10 shows the average response time to emergencies when the number of nodes varies from 250 to 500. We can see while the adaptive algorithm takes around 4 and 14 h (for $N = 250$ and 500, respectively), the static approach takes much longer time (around 24 and 45 h). The reason of such improvement is that we have differentiated emergency and normal recharging operations. However, in [5], nodes are treated regardless of their lifetime. This would incur extended recharge delay for some emergent nodes and cause battery depletion. As a result, the static approach degrades fast as the network size increases.

Fig. 5.9 Comparing the percentage of nonfunctional nodes between different schemes

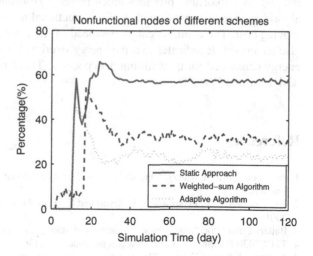

Fig. 5.10 Comparing emergency response time between different schemes

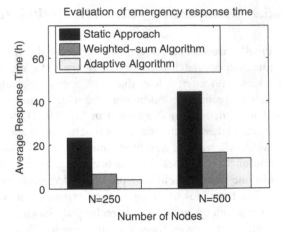

5.10   Summary

This chapter provides the performance evaluation results of some important performance metrics in WRSNs. From comparisons between the presented algorithms and optimal solution, we can see that the adaptive algorithm can achieve very close approximation to the optimal solution and provide an additional 10 % cost saving over the weighted sum algorithm. With more sophisticated algorithm design, the adaptive algorithm provides much better performance than the weighted sum algorithm in reducing the number of nonfunctional nodes, data collection delay and moving cost. The communication overhead of the node status reporting protocol is also examined. It indicates that the energy overhead is negligible compared to the energy consumed for transmitting data packets. These results can offer insights for designing real WRSNs.

References

1. Panasonic Ni-MH battery handbook, "http://www2.renovaar.ee/userfiles/Panasonic_Ni-MH_Handbook.pdf".
2. T.H. Cormen, C. E. Leiserson, R. L. Rivest and C. Stein, *Introduction to Algorithms*, MIT Press, 2001.
3. Battery Calculator, "http://www.evsource.com/battery_calculator.php."
4. TI CC2430 Datasheet, "http://www.ti.com/product/cc2430."
5. M. Zhao, J. Li and Y. Yang, "Joint mobile energy replenishment and data gathering in wireless rechargeable sensor networks," *IEEE Transactions on Mobile Computing*, vol. 13, no. 12, pp. 2689–2705, 2014.

Chapter 6
Conclusions

In this book, we discuss how to apply the novel wireless charging technology to traditional wireless sensor networks to provide perpetual network operation. First, we provide a comprehensive literature review on the state-of-the-art wireless charging techniques and their impacts. Then we introduce the network architecture for wireless rechargeable sensor networks by describing the functionality of network components and their features. We analyze several important theoretical aspects and derive important principles for the perpetual operation condition to hold.

We provide a distributed communication protocol for collecting node status in a scalable manner. We further examine the recharge scheduling problem for emergency and normal node recharge. To recharge multiple emergent nodes, we show how to formulate the problem into an Orienteering problem and describe a solution based on dynamic programming. For normal recharge, we give two scheduling algorithms. The first algorithm uses a weighted value to account for moving cost and node's lifetime. The second algorithm partitions the network adaptively, forms Capacitated Minimum Spanning Trees for each charging vehicle, and finally improves the recharge route. We also present extensive simulation results to compare the performance between the two algorithms in different criteria.

With this book, we hope to shed some light on the current research status of wireless rechargeable sensor networks, and those characterizations of various designs in the book could inspire future research. Another aim of this book is to provide a solution to tackle the energy issues in many distributed systems. We expect readers to find this book useful and supportive when facing many challenges in designing future wireless sensor networks and distributed systems.

© The Author(s) 2015
Y. Yang, C. Wang, *Wireless Rechargeable Sensor Networks*, SpringerBriefs
in Electrical and Computer Engineering, DOI 10.1007/978-3-319-17656-7_6

Glossary

© The Author(s) 2015
Y. Yang, C. Wang, *Wireless Rechargeable Sensor Networks*, SpringerBriefs
in Electrical and Computer Engineering, DOI 10.1007/978-3-319-17656-7

Printed in the United States
By Bookmasters